Building Measurement

Building Measurement

A. D. PACKER

PEARSON
Longman

Harlow, England • London • New York • Boston • San Francisco • Toronto
Sydney • Tokyo • Singapore • Hong Kong • Seoul • Taipei • New Delhi
Cape Town • Madrid • Mexico City • Amsterdam • Munich • Paris • Milan

Pearson Education Limited
Edinburgh Gate
Harlow
Essex CM20 2JE
England

and Associated Companies throughout the world

Visit us on the World Wide Web at:
http://www.pearsoned.co.uk

First published 1996

British Library Cataloguing in Publication Data
A catalogue entry for this title is available from the British
Library

ISBN 0-582-09816-5

10 9 8 7 6 5 4
07 06 05 04 03

Set by 4 in 9/11pt Compugraphic Times
Printed and bound by Antony Rowe Ltd, Eastbourne

Contents

Preface

The principal reason for writing this book was my increasing dissatisfaction with the currently available measurement texts. Most had been adapted to reflect revisions to the Standard Method of Measurements but, despite this, remained firmly based upon examples of technology that had long since been superseded. In addition, the large-scale adoption of information and communication technologies had meant that well-established practices, some only a few years old, were no longer viable or justifiable. Nevertheless, any effort at explanation which excluded a sound grasp of basic principles would be of limited value and, although technology is increasingly able to eliminate the more repetitive and tedious tasks involved in producing Bills of Quantities, it is still important to understand the composition, structure and application of the procedures involved.

The early chapters provide a context for measurement and document preparation, and consider the impact that computer and communication technology has and will continue to have on the process of Bill production. The traditional procedures associated with taking-off quantities are then described. Each of the following chapters use, wherever possible, a practical step-by-step approach to explain and interpret the Work Sections of SMM7; frequently, this information was much easier to convey by including sketches, diagrams and tables. In addition to the annotated examples which follow each chapter, extracts from dimension sheets and Bills of Quantities have been included as part of the text to illustrate presentation and approach.

Since this book is intended as an introductory text, every effort has been made to avoid unnecessary complexity. The examples that follow each chapter are therefore based on conventional domestic construction. However, once basic techniques and a logical sequence are understood and established, a similar approach can be applied to more complex situations. The confidence to do this will only come with practice.

Although *Building Measurement* is intended primarily for degree, GNVQ, diploma and certificate students taking quantity surveying, construction management and building courses, it is hoped that the book will also prove a useful reference for all those involved in the management and administration of construction work.

Despite the rigours of the Standard Method of Measurement, it is not unusual for custom and practice to vary between individual surveyors and from one office to another. With this in mind, the reader is advised that the techniques described here need not be taken as exclusive; wherever possible, a best-practice approach has been adopted.

A.D. Packer
January 1996

Acknowledgements

The author and publishers would like to thank the Royal Institution of Chartered Surveyors for permission to reproduce copyright material (our Fig. 4.1 from *Co-ordinated project information for building works, a guide with examples* published by the CCPI), and the Joint Standing Committee for Building Project Information for permission to use their examples.

There have been many people who have assisted in the preparation of this text; to all of them I am extremely grateful.

Sue Taylor and Lurleen Tamblyn for the typing of early drafts. Ruth Pearson for the computer-generated graphics. Peter Burns, FRICS, and the staff of the Southsea branch of Currie and Brown for their valued advice and help. The Editorial staff at Longman for making it happen. My students, past and present, who have suffered many of the examples adopted for this text.

My wife Linda and my children, Joseph and Samuel, who could reasonably have expected more of my time, and without whose support and understanding this text would not have been possible.

1 Introduction

1.1 Purpose and principles

The starting point in any book of this nature needs to identify to the reader its purpose and intention. To the uninitiated, the phrase 'building measurement' brings to mind the fixed notion of a tape, a building and some numbers. One thing it probably does not immediately imply is cost and forecasting. Yet the purpose of measurement in this context is inextricably linked with providing an assessment of the cost of a building, long before work has commenced on site. Initially, and most importantly, someone requires the provision of a new building. In usual circumstances they are likely to approach an architect so that their ideas and intentions can be set down on paper. It is very likely, even at this early stage, that they will need to know how much the proposals are going to cost.

Figure 1.1 A building is proposed — what will it cost?

Armed with a set of drawings, a scale rule and a calculator, measurements can be taken from these drawings and a document produced. This document identifies in some detail the component parts of the proposed works together with their quantity and will allow construction costs to be allocated to the appropriate parts of the building. Having costed each component, a forecast for the scheme can be established.

Figure 1.2 A forecast for the proposed scheme is prepared.

To enable this forecast of cost to be made with any confidence a number of basic principles must be in place. It is very important that none of the building operations are overlooked and that the items which have to be costed are presented in a recognisable form. Clearly the consequence of an error in measurement or ambiguity in a description could result in the client being ill advised with regard to the eventual cost of building operations.

Having identified the purpose of measuring building work, it is necessary to establish the general principles that will ultimately result in a document which is mutually understood and conveys the scale and extent of the construction. A common approach is necessary in terms of both

Figure 1.3 A clearly understood document conveying the scale and extent of the proposed scheme must be prepared.

Figure 1.4 Co-ordinated project information (Cpi): all of these documents are consistent and mutually interchangeable.

presenting this finished document and setting down dimensions so that others are able to understand the specification. It is not hard to imagine the confusion that would result if everyone adopted their own set of rules when measuring building work. It was exactly this situation which prompted the publication in 1922 of the first nationally recognised set of rules for the measurement of building work. An indication of the chaotic state of affairs which prevailed prior to this publication can be gleaned from reading the preface of this very first edition. Phrases such as 'diversity of practice' and 'idiosyncracies of individual surveyors' sustain a picture of confusion and doubt for the hapless early twentieth-century contractor. Some seventy years on this same set of rules (with many revisions) provides the basic principles for the measurement of building work.

Measurement can therefore be identified as the starting point from which construction costs are established. There is a standard format for the presentation of measured work and a set of rules which are mutually known and accepted. These rules are embodied in a document called 'The Standard Method of Measurement' which is currently in its seventh edition and is generally recognised by the acronym SMM7.

In turn, measurement provides the basis for the preparation of a Bill of Quantities. This document sets out the quality and quantity of all the component parts necessary for the construction of the works. It is prepared in a predetermined order which will normally follow the same sequence of work sections as SMM7.

Each tendering contractor will receive an identical set of documents including a Bill of Quantities, drawings and a specification. If the Common Arrangement of Work Sections has been adopted (see below), all of these documents will cross-reference to each other.

The costing columns of the Bills of Quantities are completed, extended and totalled independently by each tendering contractor to establish a tender price for the completion of the construction work. Individual tenders are submitted and compared with the other tenders received. The client, acting on advice, normally accepts the most suitable tender and enters into a formal agreement with the selected contractor. The Bills of Quantities, along with the other tender documents, become part of the contract documentation to which both parties are formally contracted. Whilst this fulfils the principal function of the Bill of Quantities (and thereby the measurement process), it also provides a valuable role in the financial management of the project during, and to some extent after, the construction process. There are many sophisticated variants of the tender procedure described above, but whichever technique is used, all construction costs will, at some stage, have been prepared from quantities established by the process of measurement.

The remainder of this text identifies, largely by example, the techniques that are commonly practised in drafting Bills of Quantities prepared in accordance with SMM7. (Custom and practice may well vary between individual surveyors and the procedures given should not be considered as irrevocable.) Despite the rigours of a Standard Method of Measurement, it is not unusual for individual practices to vary considerably. With this in mind, the reader is advised that the techniques practised here need not be taken as exclusive.

1.2 The classification system

In the very first place someone has to agree a system of

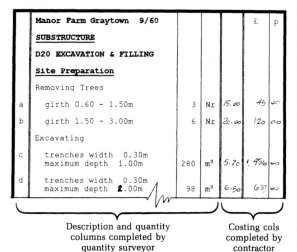

	Manor Farm Graytown 9/60				£	p
	SUBSTRUCTURE					
	D20 EXCAVATION & FILLING					
	Site Preparation					
	Removing Trees					
a	girth 0.60 - 1.50m	3	Nr	15.00	45	00
b	girth 1.50 - 3.00m	6	Nr	20.00	120	00
	Excavating					
c	trenches width 0.30m maximum depth 1.00m	280	m³	5.20	1,456	00
d	trenches width 0.30m maximum depth 2.00m	98	m³	6.50	637	00

Description and quantity columns completed by quantity surveyor

Costing cols completed by contractor

Figure 1.5 An example of a page from a Bill of Quantities showing the description and quantity columns (prepared by the quantity surveyor) and the costing columns (handwritten − completed by the contractor's estimator at the tender stage).

classification for the construction process. It has to be robust enough to embrace the variety of trades employed in the construction process, detailed enough to allow for technical distinctions and commonly understood by all those who use it. The Common Arrangement of Work Sections (CAWS) was established to achieve these goals and was adopted as the framework within which the rules for measuring building work under SMM7 were to be drafted. The classification system is loosely based on the pattern of trades employed during building operations. The order in which these are presented generally reflects the sequence of events as they are likely to occur on site. Whilst they might not always be obvious, the majority of these Work Sections are clearly identifiable from their titles.

For the purposes of measurement, SMM7 identifies a set of rules under each of these Work Sections. These are presented in a tabular form using two levels of heading; the first is identified by the letters of the alphabet (see figure 1.6) and the second by numerals. Whilst a finer grain of definition is available, it is not used by SMM7. A full listing of the classification showing both level 1 and level 2 headings is given in the Detailed Contents of SMM7.

The adoption of this Common Arrangement of Work Sections has meant that when Bills of Quantities are prepared using SMM7 they should cross-reference to any other documents (such as drawings and specifications) which have been prepared using the same system of classification. The hope is that when correctly implemented, the contractor will receive a set of site documents which are consistent and mutually interchangeable.

Whilst specific rules relating to particular trades can be

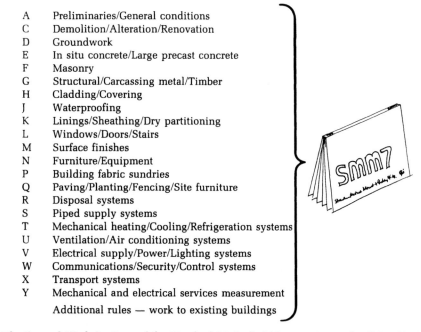

A	Preliminaries/General conditions
C	Demolition/Alteration/Renovation
D	Groundwork
E	In situ concrete/Large precast concrete
F	Masonry
G	Structural/Carcassing metal/Timber
H	Cladding/Covering
J	Waterproofing
K	Linings/Sheathing/Dry partitioning
L	Windows/Doors/Stairs
M	Surface finishes
N	Furniture/Equipment
P	Building fabric sundries
Q	Paving/Planting/Fencing/Site furniture
R	Disposal systems
S	Piped supply systems
T	Mechanical heating/Cooling/Refrigeration systems
U	Ventilation/Air conditioning systems
V	Electrical supply/Power/Lighting systems
W	Communications/Security/Control systems
X	Transport systems
Y	Mechanical and electrical services measurement
	Additional rules − work to existing buildings

Figure 1.6 The General Work Sections of the Standard Method of Measurement, 7th edition (SMM7).

found in SMM7 under the appropriate work section, a set of general rules precedes these and applies to the measurement of all construction operations. These 'provide a uniform basis' for measuring, describing and billing building works. To this end they identify the standard of accuracy necessary for recording quantities (General Rules 3.1−3.5) together with definitions and interpretations for written descriptions (General Rules 4.1−4.7). These techniques are considered in more detail in the following chapter.

2 Traditional BQ preparation

2.1 Taking-off quantities

2.1.1 Introduction

In the first instance quantities will need to be extracted from the drawings, together with an appropriate description. This process, known as booking dimensions or taking-off quantities, involves the measurer in either reading or scaling dimensions from the drawings. There are two distinct parts to this. The first involves the recording of quantities, whilst the second requires a written description to accompany the quantity. The sequence adopted by measurers in this initial stage bears little relation to the eventual order of the finished Bill of Quantities. This is because 'taking-off' has been devised in order to assist the measurer with both the speed and accuracy of recording dimensions and largely follows the sequence of events as they will occur on site. At a later stage these quantities will be arranged in the sequence of SMM7. For example, when measuring a foundation trench the excavation, disposal, concrete work and masonry are all measured at the same time, regardless of the eventual location of these items in the finished Bill of Quantities. This particular pattern of measured items is generally known as the *Group Method* since it reflects a common set of dimensions that any number of different trades might share. Grouping of items is therefore determined not by their eventual position in the BQ but by their dimensions.

The alternative to the group method of measurement is known as the *Northern Method* or the trade-by-trade approach. As its name suggests each item should be measured in trade sequence (or the order dictated by the Standard Method). This means scanning all the drawings to determine the order of measurement and allows little opportunity for the grouping together of items with a common set of dimensions. Whilst this approach is often used for small projects, it is inappropriate for larger, more complex work.

Whilst there will always be a need to understand the mechanism by which Bills of Quantities are produced, it is unlikely that the present generation of quantity surveyors would undertake the task of bill preparation without the assistance of a computer. The technology commonly available allows dimensions to be entered via a digitiser which is linked to a library of descriptions held in a database. These quantities are automatically located and sorted into eventual bill order and the final document may be presented in any number of styles or formats. It is claimed that this technology can reduce bill preparation time by up to two-thirds. It is quite possible that the technology which is able to generate an architect's intentions in three dimensions may also automatically produce Bills of Quantity. The impact of these developments is considered further in chapter 4. The remainder of this chapter assumes the conventional approach using the group method of measurement.

Once the take-off is complete the measured items will be squared and totalled. All of the arithmetic should at this stage be checked.

2.1.2 Setting down dimensions

It is important to appreciate that the techniques adopted in the preparation of a Bill of Quantities will vary from one individual to another and between one office and the next. For the most part it is assumed that the operation of transferring drawings into the descriptions and quantities that are collectively termed a 'Bill of Quantities' will be carried out without the assistance of a computer. Clearly, there are significant advantages in using a computer package that can generate digitised quantities and automatically sort measured items into the appropriate Work Section. Nevertheless, in order to fully understand the detail of this process, it is necessary to examine each step in turn and this cannot be fully appreciated by simply pressing a button. What follows is therefore based on a procedure which has

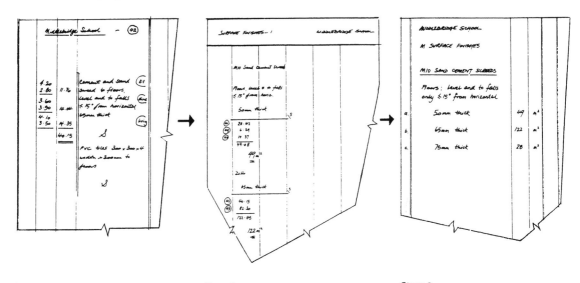

Stage 1
Quantities are prepared on dimension paper and involve the transfer of drawn details into descriptions and dimensions. This process is often termed 'booking dimensions' or 'taking-off quantities'.

Stage 2
These squared and totalled quantities are arranged in a predetermined sequence (normally SMM7 work sections) and identical items are amalgamated. This process is called 'abstracting' or 'working-up quantities'

Stage 3
The final stage. Descriptions and quantities are prepared in draft form ready for typing. The finished master document is 'read-over' before being photocopied and sent with other tender documents to tendering contractors.

Figure 2.1   The stages in the preparation of traditional Bill of Quantities.

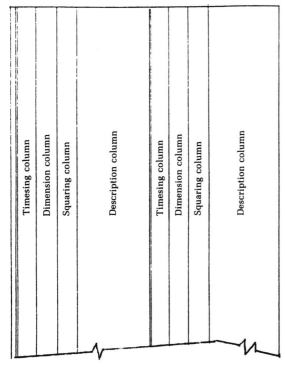

Figure 2.2   The format of standard dimension paper.

evolved over a number of years and is best described by the term 'traditional bill preparation'.

Dimensions are taken from a drawing and recorded on specially lined paper known as *dimension paper*.

The A4 page is divided vertically into two identical halves each comprising a set of four columns. These are labelled above for the purpose of identification. The extra column on the extreme left is called the binding margin and would not normally be used for recording dimensions.

Dimensions are read directly from the drawings and recorded to two decimal places of a metre in the vertical column labelled DIMENSION COLUMN. Alternatively these dimensions might be scaled directly from the drawing and entered in the same way. Once these have been entered it will be necessary to provide some form of description. The widest of the four columns, labelled the DESCRIPTION COLUMN, is used for this purpose.

When quantities appear in the Bill of Quantities, they will be followed by the appropriate unit of measurement for the work (see sample Bill of Quantities). As they are being booked on a sheet of dimension paper they will appear in a number of different forms and, to the untrained eye, there would appear to be no obvious indication of a unit of measurement. It is necessary to understand these before any dimensions can be booked. The principal unit for recording dimensions is the metre and this is expressed in

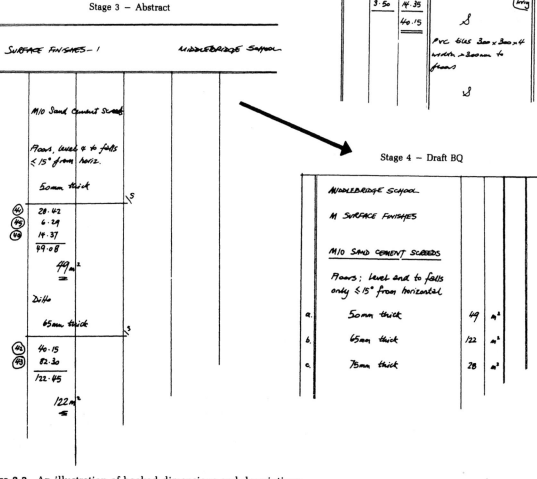

Figure 2.3 An illustration of booked dimensions and descriptions.

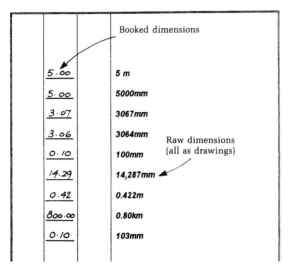

Figure 2.4 Recording dimensions. Often architects and engineers will use different forms of metric dimensions and it is the measurer's task to record these in the dimension column to **two decimal places of a metre**, i.e. a dimension of 5 mm or over is taken as 10 mm; less than 5 mm is disregarded.

the dimension column to two decimal places, i.e. the nearest 10 mm (SMM7 General Rule 3.2). Accordingly, 3067 mm would be recorded in the dimension column as 3.07. There is no need to write the word metre or use the letter m, since all dimensions are assumed to be recorded in the same way. Even when the dimension is a whole unit (5 metres), two zeros should be used after the decimal place, as in the example of figure 2.4.

Before describing the purpose of the remaining two columns, it would be prudent to spend some time looking at the various ways in which dimensions are recorded. The technical term for entering dimensions in this way is 'booking dimensions' and these should always be recorded to *two decimal places of a metre* (SMM7 General Rule 3.2).

The basic component of measurement is the *item* and it will normally have three parts:

• Quantity
• Description
• Unit of measurement

The *quantity* is the arithmetical result of booked dimensions, the *description* is the written explanation of what is being measured, whilst the *unit* may be any one of the following:

nr.   (indicating 'enumeration', i.e. items counted)
m   (metre — linear measurement, i.e. length)
m²   (square metre — superficial measurement, i.e. area)
m³   (cubic metre — volume)
Item (description only — no quantity)

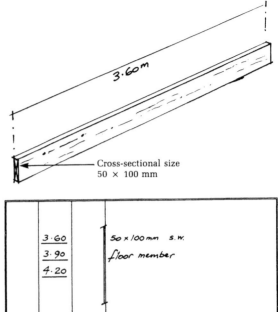

Figure 2.5 Linear metres. Two further joists have been booked in this example (both of identical cross-sectional size) and grouped with the 'common description': one at 3.90 m in length, the other at 4.20 m. A line beneath each dimension identifies the unit of measurement as a length.

The following two units are less frequently used in SMM7:

t   (tonne — measurement by weight)
h   (hour — expression of time)

The four principal units of measurement are *enumeration*, *length*, *area* and *volume*. In the previous example (figure 2.4) all of the booked dimensions were recorded in linear metres. It would be inappropriate to use this same unit when measuring the excavation of a foundation trench which has not only a length but, in addition, a width and a depth. It may have been noted in the previous example that a line was drawn across the dimension column under each set of recorded dimensions. This identifies each single entry as an individual length. The technical term for this unit of measurement is *linear metres*.

The principal unit of measurement associated with excavation is *cubic metres* and this is identified in the dimension column by *three dimensions* set down one above the other and grouped together by a horizontal line across the dimension column under the last dimension.

In similar fashion, dimensions grouped together in pairs are automatically associated with items that have been measured in *square metres*, such as brickwork and plasterwork.

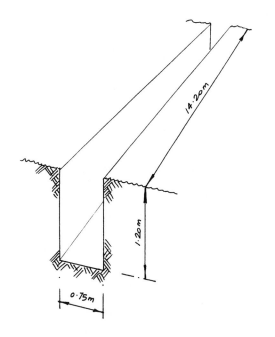

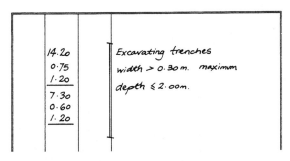

Figure 2.6 Cubic metres. Here two foundation trenches have been booked. Each have the same description and are recorded together in cubic metres. The longer of these two trenches has been sketched.

In some cases it is difficult to identify an appropriate unit of measurement and in these instances counting or *enumeration* is used. Door furniture, manhole covers and lintels are all measured using enumeration. These appear as whole numbers in the dimension column, with a line drawn horizontally beneath each single entry.

To summarise:

- Dimensions are recorded to *two* decimal places of a metre.
- There is no need to identify the unit of measurement since this is clear from the presentation.
- Where dimensions appear in sets of two or three, these will be multiplied together (squared) to show an area or volume.

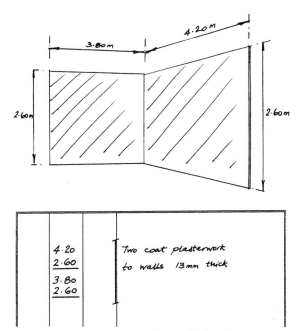

Figure 2.7 Square metres. Two walls have been plastered in this case. Both have a height of 2.60 m and share a common specification. One is 4.20 m long, the other 3.80 m. The unit of measurement is square metres, indicated by a line below each pair of dimensions.

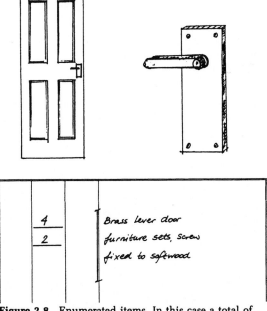

Figure 2.8 Enumerated items. In this case a total of six pairs of door handles have been booked, four in one location and two in another. A whole number appears in the dimension column with a horizontal line drawn across the column below each entry.

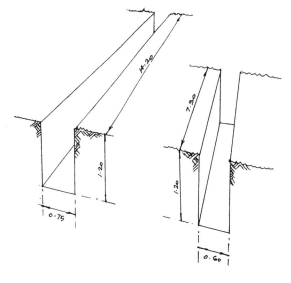

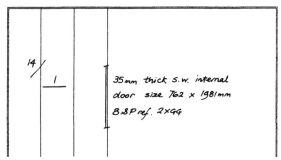

Figure 2.10   Timesing. Here 14 identical doors are found on the drawings and the timesing column is used to record this. The slash line indicates that the quantity should be multiplied by the number above it.

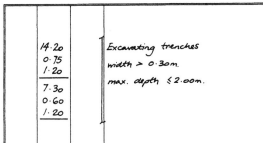

Figure 2.9   The sequence and presentation of booked dimensions. A simple scan shows two different lengths and widths of trench with a consistent depth of 1.20 m.

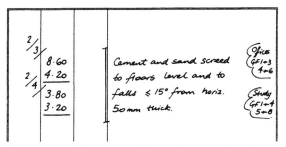

Figure 2.11   Timesing can be used any number of times to indicate, for example, an identical set of rooms on two storeys.

- Clear, legible presentation, in ink, with items well spaced.
- The order of recording dimensions is
  (1) length
  (2) width or breadth
  (3) vertical height or depth.

This last point is of no significance to items of length or enumeration, but is of great value when tracing the build-up of dimensions for areas and volumes.

2.1.3 The timesing column

So far we have only considered using the dimension column and the description column.

The term 'timesing' is used by surveyors to identify the number of repeats a quantity may have in the finished building. For example, when measuring a number of identical internal doors, it is convenient to simply record dimension

and description only once, and then 'times' this by the number of repeats. A great deal of time can be saved during measurement by the correct use of the timesing column. It may also go some way towards providing an indication of the layout, simply from the dimensions.

2.1.4 Dotting-on

Where identical dimensions are repeated it may be necessary to add rather than to multiply. For example, six pits of identical size are to be excavated and their dimensions are recorded as in figure 2.12. Subsequently, an additional two pits are spotted and rather than enter another set of dimensions together with an identical description, the original set can be adapted by 'dotting-on' another two as in figure 2.13. This technique is also carried out in the timesing column and, technically, any number of additions can be made, although this will be somewhat restricted by the limitations of space and the need for legible presentation. The dot will usually be placed below the original figure to allow more space, whilst the timesing slash will be placed above. In practice these two techniques are often combined and this is illustrated in figures 2.14 and 2.15.

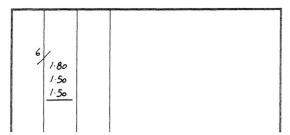

Figure 2.12   Here six identical pits have been recorded.

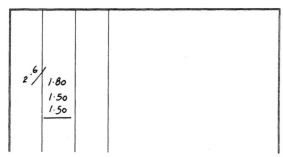

Figure 2.13   A further two pits are added by dotting-on.

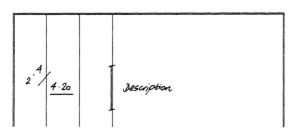

Figure 2.14   Combining timesing and dotting-on. Dimensions show a multiplier of six: total length = 4.20 m × (4 + 2).

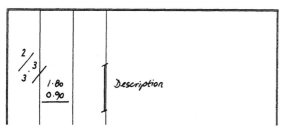

Figure 2.15   Another combination of timesing and dotting-on. Dimensions show a multiplier of 12: total area = 1.80 × 0.90 × (2 × (3 + 3)).

2.1.5 Geometric forms

Frequently it is necessary to book dimensions for triangles, circles and other irregular figures. This can be achieved by a combination of figures in both the dimension and

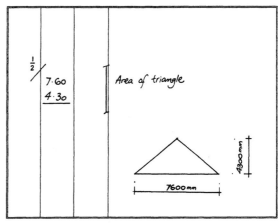

Figure 2.16   The area of a triangle.

timesing columns. The important point to remember is the unit of measurement. Most geometric formulae are quite easily transposed into booked dimensions. For example, the formula for the area of a triangle is half the base multiplied by the height. This can be recorded on dimension paper as in figure 2.16. The area of a circle of radius 8.64 metres (formula πr^2) can be recorded as in figure 2.17, and of a sector as in figure 2.18. If, on the other hand, we needed to record the circumference of this circle (formula $2\pi r$), only a single dimension would appear in the dimension column, as in figure 2.19. The inclusion of a fraction in combination with the above can give the circumference length of any portion of the circle (e.g. a semi-circle).

The formulae for regular geometric forms are given at the end of this book.

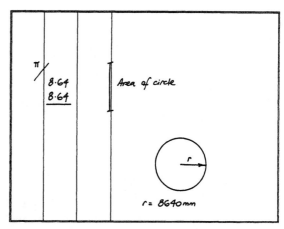

Figure 2.17   The area of a circle.

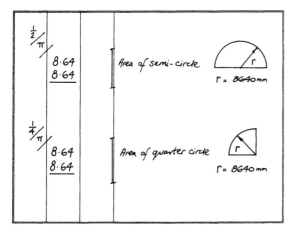

Figure 2.18 The area of a sector of a circle.

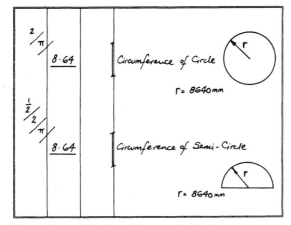

Figure 2.19 The circumference of a circle.

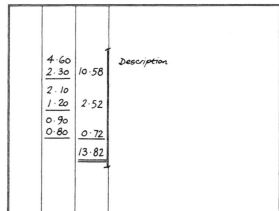

Figure 2.20 Squaring quantities. Frequently these sets of dimensions will appear in groups and in this case each individual result is recorded in the squaring column, parallel with the last dimension of the particular set. These results are in turn tallied to provide the total for this particular item of measurement. Past experience has shown that students find it difficult to resist the temptation to square quantities as they are recorded. This practice should be avoided since all the dimensions and descriptions must be completed before squaring can commence.

2.1.6 The squaring column

Attention should now be turned to the function of the squaring column. As has been previously noted, dimension figures are recorded in the dimension column to indicate the unit of measurement. In the case of area and volume, dimensions will be multiplied together and the result of this computation is entered in the squaring column. There is never any need to include a multiplication sign since this is assumed.

2.1.7 The description column

As its name suggests, this is where the worded part of the measurement is recorded. The order and form of wording is important since it should convey concisely all the information necessary to allow a price to be established. The introduction of the tabulated arrangement of SMM7 has,

to a large extent, provided a framework around which to build descriptions. Despite this, there still remains a degree of flexibility in the interpretation and presentation of written descriptions. So far as possible, the structure and terminology of SMM7 together with the philosophy of the Common Arrangement has been adopted in the preparation of this text.

This is undoubtedly one of the more difficult parts of booking dimensions and the novice is advised to spend some time looking through a Bill of Quantities before attempting to write descriptions for measured work (figure 2.21).

2.1.8 Waste calculations

Ideally dimensions can be read directly from the drawings and entered to two decimal places of a metre on dimension paper. Frequently this direct transfer is not possible since some adjustment is required to the dimensions before they can be booked. These preliminary calculations are known as waste calculations or side-casts and are presented to the nearest millimetre on the right-hand side of the description column immediately above the item to which they relate. Once the required dimension is identified by waste calculation, it is reduced to two decimal places and transferred to the dimension column. There may be a temptation to scribble these in note form or even carry out simple arithmetic in the head. Both should be avoided since it is

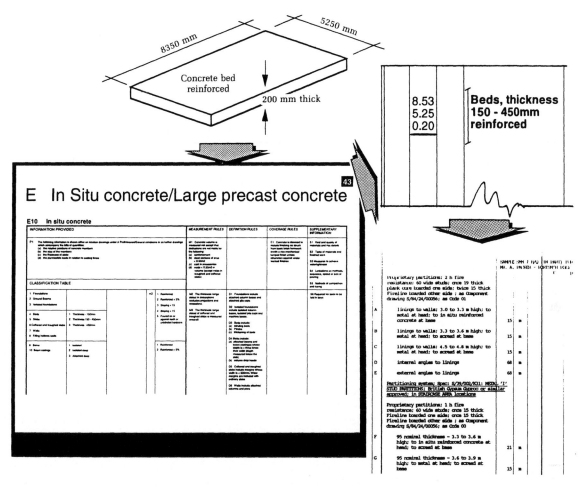

Figure 2.21 Interpreting SMM7, drawn details and writing descriptions.

important to identify the process by which the dimension was established. Double underlining in a waste calculation usually indicates that the result has been transferred to the dimension column (figure 2.22).

2.1.9 Bracket lines, the ampersand (&) and the group method of measurement

Frequently more than one set of dimensions relate to the same description and when this occurs it is necessary to link together both description and dimensions with a *bracket line* (figures 2.23 and 2.24).

On occasions it is convenient to use an ampersand to link two or more items that have different units of measurement. This is perhaps best demonstrated by the example shown in figure 2.25. This is a more advanced technique which

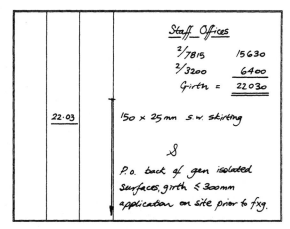

Figure 2.22 Setting down waste calculations.

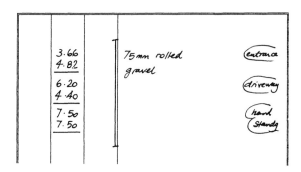

Figure 2.23 Bracket lines. This vertical line is drawn just inside the description column and extends to embrace all the dimensions which relate to that description. The top and bottom of the bracket line is marked with a short horizontal line. The group method of measurement presupposes that descriptions with shared dimensions appear together in the take-off and a bracket line is used to show this linkage.

the student is advised to avoid until such time as the basic principles of booking dimensions are clearly understood.

2.1.10 Spacing of dimensions and signposting

The spacing of dimensions and descriptions is an important part of the measurement process. Clear, well-spaced dimensions are easy to follow and can be readily checked by others. Whilst there are no written rules about the presentation of measured items, the extract in figure 2.26 provides

an indication of the spacing and layout of a typical sheet of booked dimensions.

Every effort should be made in the take-off to ensure that dimensions and waste calculations can be traced back to the drawing. Signposts or location notes can be used in the description column to provide this cross-reference. These should appear on the right of the description column parallel with the dimension to which they relate. A simple line or ring around this location note prevents it being read as part of the description.

2.1.11 Abbreviations

The use of abbreviations when writing descriptions is commonly practised by measurers. Apart from the practical limitations of space a great deal of time is saved by shortening the more frequently used words. A full list of the more commonly recognised abbreviations is given at the end of this book. Individual surveyors and practices tend to develop their own forms of abbreviations, and whilst there are no hard and fast rules it is always important to bear in mind that others must be able to fully understand the written description. Whilst abbreviations are acceptable in both the take-off and the abstract, their place in the completed Bill of Quantities should be restricted to those listed in SMM7 General Rule 12.1. Should there be the slightest chance of a word or term being misunderstood in the finished Bill of Quantities, it must be written in full.

Many of the descriptions in a take-off are repeated several times, often in the space of a few pages. Where this occurs,

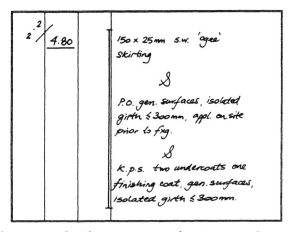

Figure 2.24 Anding-on (the ampersand). Where two or more descriptions apply to a single set of dimensions, each description is separated by an ampersand. This process of 'anding-on' is fundamental to the group method of measurement and avoids the need to write down dimensions more than once. Care should be taken to ensure that the same unit of measurement applies to each of the items linked together by the ampersand, since the total of all the quantities on the left of the bracket relates to **each and every** description on the right of the bracket.

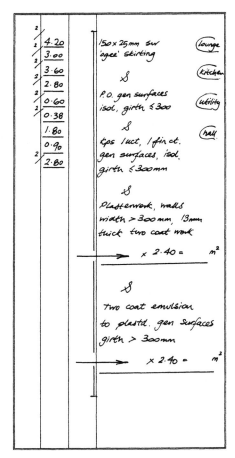

Figure 2.25 Constant dimension. In this case three different classes of work have been grouped together since they share a common set of base dimensions, i.e. the girth or perimeter length of a room. The first three items are measured in linear metres, whilst the last two are measured as areas. Rather than starting a new set of dimensions, it is far simpler to introduce a conversion factor (in this case the constant floor-to-ceiling height). This must be made clear to the person squaring and is usually shown by arrows leading across from the squaring column into the description column where the conversion takes place. A space for the resulting quantity must be left and the appropriate unit of measurement should be identified.

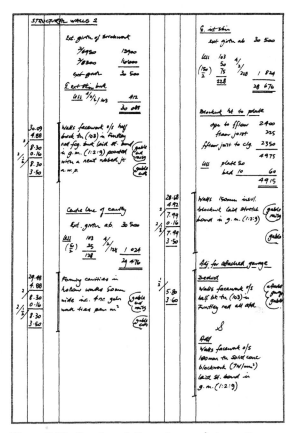

Figure 2.26 Spacing of dimensions and signposting.

a great deal of time can be saved by referring to a previous, similar or identical description. This is achieved by using the abbreviation *abd* (as before described), as in figure 2.27.

The word 'ditto' is often used in the same way, although in this instance it is understood to refer to the immediately previous description. For example, figure 2.28 shows the dimensions which might be booked when measuring drainage work.

Care should be taken when using either technique since it is easy to misdirect the intended back-reference.

2.1.12 Adjustments

This is the term that measurers use to describe an alteration to a set of previously recorded dimensions. Generally this is necessary because it is easier and safer to overmeasure an item in the first instance and make a deduction or adjustment later. Consider the measurement of a carpet to an office floor which is interrupted around its perimeter by a number of concrete columns (figure 2.29). Rather than attempt to break this area down into a series of strips, it is far simpler to overmeasure initially and then deduct the area of flooring covered by the columns.

It is usually assumed that all booked items are additions and there is consequently no need to write the word '*add*' each time a description is recorded. The opposite of this is the case with respect to *deductions* and it is vital that this is made clear to the squarer. A deduction will generally follow the item which gave rise to that adjustment.

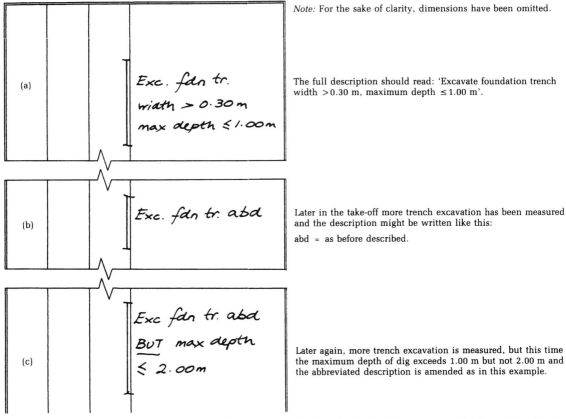

Note: For the sake of clarity, dimensions have been omitted.

(a) Exc. fdn tr. width >0·30 m max depth ≤1·00m

The full description should read: 'Excavate foundation trench width >0.30 m, maximum depth ≤1.00 m'.

(b) Exc. fdn tr. abd

Later in the take-off more trench excavation has been measured and the description might be written like this:

abd = as before described.

(c) Exc fdn tr. abd BUT max depth ≤ 2.00m

Later again, more trench excavation is measured, but this time the maximum depth of dig exceeds 1.00 m but not 2.00 m and the abbreviated description is amended as in this example.

Figure 2.27 Abbreviations and short written descriptions. The description in (a) appears on the dimension sheet. Later in the take-off more trench excavation is measured and the description might be written as (b) implying all the detail of the previous description. Later again more trench excavation is measured, but this time the maximum depth of it exceeds 1.00 m but not 2.00 m and this is described as (c).

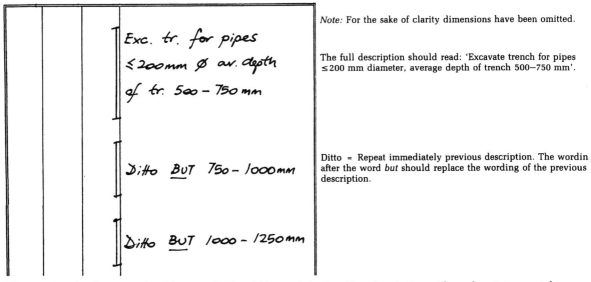

Note: For the sake of clarity dimensions have been omitted.

Exc. tr. for pipes ≤200mm Ø av. depth of tr. 500 – 750 mm

The full description should read: 'Excavate trench for pipes ≤200 mm diameter, average depth of trench 500–750 mm'.

Ditto BUT 750 – 1000mm

Ditto = Repeat immediately previous description. The wording after the word *but* should replace the wording of the previous description.

Ditto BUT 1000 – 1250mm

Figure 2.28 Another example of the use of abbreviation and short written descriptions. These descriptions might be booked when measuring drainage work.

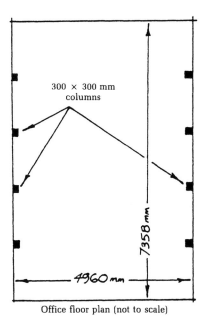

300 × 300 mm
columns

7358 mm

4960 mm

Office floor plan (not to scale)

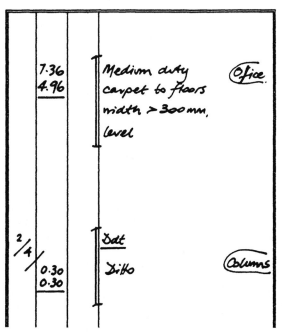

7·36	Medium duty	Ofice
4·96	carpet to floors	
	width > 300 mm,	
	level	
2/4		
0.30	Ddt	Columns
0.30	Ditto	

Figure 2.29 Adjustments. The area of the columns is deducted from wall-to-wall floor area to yield the true area of carpet.

There are two methods generally recognised for making adjustments, and their difference is purely a matter of presentation. Examples of each of these are given in figure 2.30.

Care should be taken when making adjustments to a

number of descriptions which are linked to the same quantity using an ampersand. This can become particularly confusing unless the word 'deduct' is included for every adjustment and not just the first. In figure 2.31, whilst the intention of the measurer is clear, once the first description is transferred to the abstract and lined through, the squarer has lost all reference to the instruction to deduct the remaining items, and it is likely that these quantities will be added rather than omitted.

2.1.13 Location notes/signposting

An often overlooked but nonetheless important part of taking-off quantities is the presentation of dimensions and descriptions. As dimensions are booked it is useful to make a brief note to identify the source or location of the quantities. Should there be a need to check how a Bill of Quantity item was established, the individual sets of dimensions can be traced back to the drawing. These notes are referred to as 'signposts' or 'location notes' and are recorded on the right-hand side of the description column parallel with the quantity to which they relate. To avoid their becoming confused with the description of the work being measured they should be ringed.

2.1.14 Correction of errors

Inevitably mistakes will occur in the take-off and regardless of the cause there is an established procedure for making corrections. This practice has become established since it is often important to see how and why a mistake occurred. To achieve this the corrected figure or item must remain legible, although clearly cancelled. Figures 2.33 and 2.34 illustrate two different types of correction, the first in a waste calculation and the second a late design change.

2.2 Abstracting or 'working-up' Bills of Quantities

The sequence adopted by measurers using the group method of measurement largely follows construction operations as they occur on site. However, once the take-off is complete, these measured items will need to be collated, like items must be merged and deduction adjustments made. This process, known as *abstracting* or *working-up quantities*, is carried out on specially lined A3 size paper.

At the head of each abstract a Work Section heading is recorded, together with any other references for the project. Each measured item is copied from the dimension column and transferred to the abstract in SMM7 order. In an effort to avoid double transfer, or the omission of an item, each description is lined through on the dimension sheet as it is transferred.

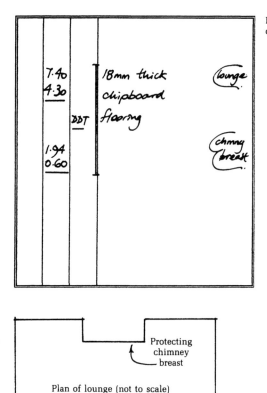

Plan of lounge (not to scale)

Here the adjustment is made as part of 'squaring' and a net quantity identified on the sheet of dimension paper.

In this instance the positive and negative quantities are presented as two separate items and the net quantity will only be established once the abstracting stage is complete.

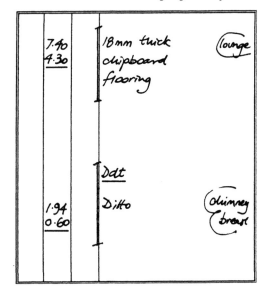

Figure 2.30 Adjustment methods: method 1 – the adjustment is made on the dimension sheet; method 2 – individual totals are squared and transferred to an abstract before any adjustment is made.

Descriptions are copied spanning two columns on the abstract and headings included to provide the framework of the final document. A horizontal line is drawn below each transferred description and the squared quantity entered below this line, additions on the left and deductions on the right. To provide a cross-reference to the dimension page, each squared quantity is labelled with its source (i.e. the dimension page number). Finally, the appropriate unit of measurement, identified by the letters L (Linear), S (Square Area), C (Cubic Volume) or Nr. (Enumerated Items), should be entered since all reference to these would otherwise be lost.

Each transferred item is entered in this fashion. Items from different parts of the take-off will be amalgamated under a single common description. Related Work Section items from different parts of the take-off will appear on the same abstract. The abstracted items should be well spaced apart, allowing for the later insertion of omitted

Table 2.1 Common arrangement of Work Sections in SMM7

Level	Work Section	Example billing
1	Main Work Section	M SURFACE FINISHES
2	Not used by SMM7	—
3	Work Section	M20 Plastered/Rendered/ Roughcast Coatings
4 5 6	Used as appropriate in accordance with classification tables of SMM7	Spec. ref. M20.ATC ceilings to timber base width > 300 mm; 12.7 mm th.

items. Once all measured items have been transferred to the abstract, the quantities are totalled and rounded to the nearest whole unit (SMM7 General Rule 3.3).

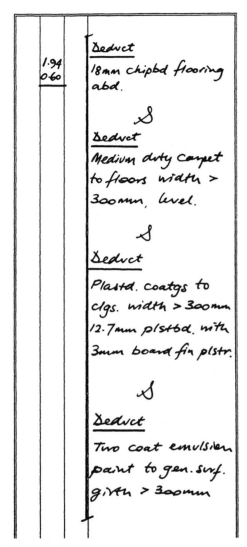

Figure 2.31 The heading 'Deduct' should be used to avoid ambiguity.

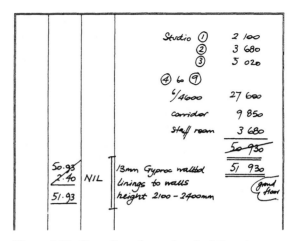

Figure 2.33 Correction of a waste calculation: 1 – amend waste with a neat cancellation and enter corrected figure below; 2 – cancel the incorrect set of dimensions with the word 'NIL' written in the squaring column and enter the corrected dimensions; 3 – extend the bracket line to include these corrected dimensions.

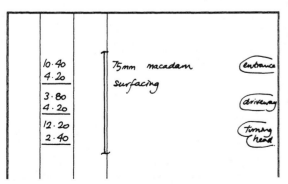

Figure 2.32 Location notes and signposts.

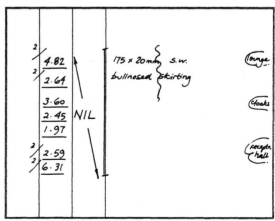

Figure 2.34 Here the architect or client decided on a change of design after the original proposals had been measured: 1 – cancel and 'NIL' the dimensions as before; 2 – cancel the description with a bold wavy line drawn vertically through the description (which will later join the straight vertical abstracting line).

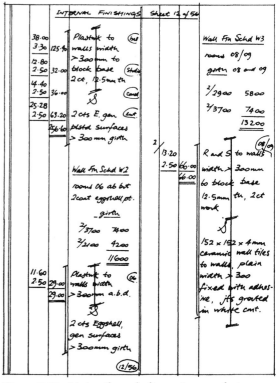

Figure 2.36 Lining through dimensions on their transferral to abstract.

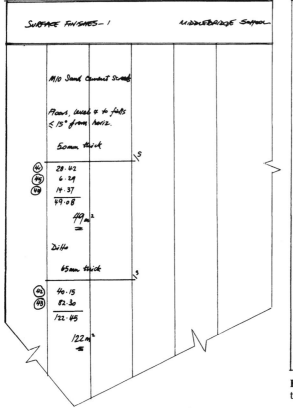

Figure 2.35 Layout and spacing of a completed abstract.

2.3 Billing

This is the final stage in the preparation of the completed Bill of Quantities. The effort of assembling and ordering was completed at the abstracting stage and all that remains is for the descriptions and quantities to be presented in a structured and consistent fashion. When SMM7 is used, this presentation will follow the Common Arrangement of Work Sections (see figure 1.6). This identifies three levels of heading, the third and lowest level being the Work Sections. The three levels have the following titles:

Level 1: Group (e.g. D Groundwork)
Level 2: Sub-group (e.g. D3 Piling)
Level 3: Work Section (e.g. D30 Cast in place concrete piling)

Because of the tabular arrangement of SMM7, further divisions within each Work Section can be defined and these are given in table 2.1 as a series of levels.

The use of upper and lower case letters together with underlining, bold face print and indenting helps to identify this priority in the completed Bill of Quantity (figure 2.37).

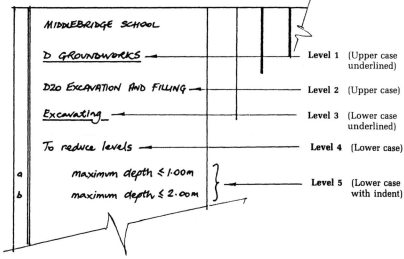

Figure 2.37 Draft Bill of Quantities with levels of heading identified.

3 Mensuration: techniques and conventions

Mensuration is the term used by mathematicians to describe the techniques used to establish lengths, areas and volumes. It is necessary to understand the principles of mensuration before dimensions can be correctly presented and recorded on dimension paper. Whilst many people are unfamiliar with the term 'mensuration', most of the geometric formulae are generally well known. This chapter provides an introduction to the more commonly adopted mensuration techniques practised by measurers. (A full list of geometric formulae is given at the end of this book.)

3.1 Girths

One of the most frequently used techniques when booking dimensions is 'girthing'. Most buildings are based on a square or rectangular plan shape and it is often necessary to establish the perimeter length of individual rooms or whole buildings either internally or externally. Whilst the drawings will show plan dimensions, before these can be set down and recorded on dimension paper it will be necessary to build up perimeter lengths as waste calculations. For example, if we wanted to measure the skirting or coving in a room, it would be convenient for the dimensions of the room to be given in the dimension column as a single linear item (i.e. the perimeter). This is carried out as a waste calculation which clearly identifies the individual plan lengths as given on the drawing. It would be incorrect to carry out this calculation in the head and simply assume everyone knew exactly what you had done. One of the basic rules of recording dimensions is that they are clearly presented and can be traced back to the drawing.

The waste calculation to establish the internal girth of the outline room plan shown in figure 3.1 may be presented either as in figure 3.2 or as in figure 3.3. Both show the dimensions as they appear on the drawing (allowing anyone else to check them against the drawing). External girths are established in the same way.

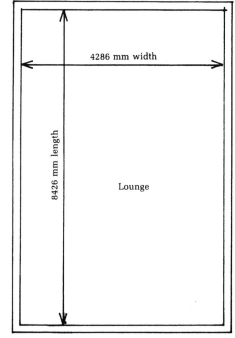

Figure 3.1 Plan of a room layout.

In the example above (figures 3.1, 3.2 and 3.3), the girth of the room was used to establish the length of the skirting. It might also have been used as the basic dimension to establish the internal wall area.

In the example shown in figure 3.4, the unit of measurement is square metres, but the booked dimensions are based on the room's girth. If the girth is used in combination with the floor to ceiling height, it will provide the internal wall area. In this case it is applicable to two items, wall plaster and emulsion paint, and both are linked to their common dimensions by an ampersand and bracket line.

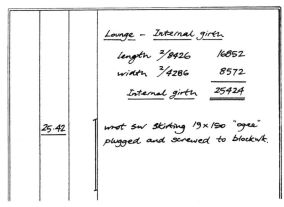

Figure 3.2 Internal girth — first representation.

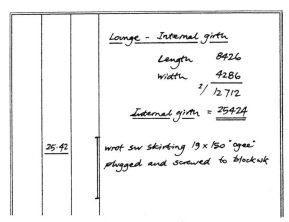

Figure 3.3 Internal girth — second representation.

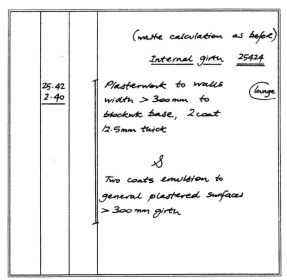

Figure 3.4 Establishing the internal wall area.

If the external girth is required and the drawn details only show internal dimensions, adding the wall thickness to each plan dimension (once at each end) will give the external plan dimensions. Using the same girthing technique as before, the external perimeter length can be presented as a waste calculation (figure 3.5).

3.2 Centre lines

When measuring brickwork or trench excavation it is necessary to book dimensions based on average or mean lengths. This is achieved by an adaptation of girthing, known as a *centre line calculation*. Rather than measuring some items as surface areas based on a girth (such as plasterwork and decoration to walls) it is occasionally more accurate to base the measurement on the mid-point of the material being recorded. This is perhaps best demonstrated by taking the example of trench excavation. Most foundation trenches are 450 mm or 600 mm wide; in order to measure them accurately it will be necessary to base all dimensions on the centre line length. In a situation where we are presented with internal dimensions an internal girth can be found. If the internal plan dimensions are adjusted by adding half the total trench width at both ends, the centre line is established.

Rather than thinking of the technique in terms of adding to individual plan dimensions, it is more appropriate to consider the effect of this adjustment at each corner. This is only because it helps to explain the presentation of the waste calculation. For the four corners of a regular plan-shaped building, the equation is simply timesed by four, which in this case (since internal dimensions were used) means the result is added to the girth to give the centre line. Had an external girth been used the same result could be achieved by deduction to provide the centre line (figures 3.6 and 3.7).

3.3 Irregular plan shapes

So far we have only considered buildings which are regular in plan shape. In many cases buildings are designed with clipped corners and insets and these must be taken into consideration when establishing girths and centre lines.

In the case of a clipped corner, so long as the angles remain at 90°, there is no need for any adjustment in either the girth of the building or its centre line.

In figure 3.9 the solid line shows the actual plan outline of the building and the hatched outline the line of a regular-shaped building. It can be seen that the corner could be 'folded out' to form a regular right-angled corner. In any girthing exercise the clipped corner has no effect, since an internal angle compensates for an external angle. This rule of compensation may be applied whatever the shape of a

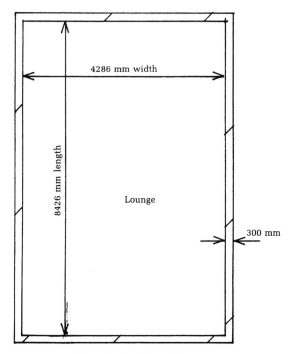

Floor plan (not to scale)

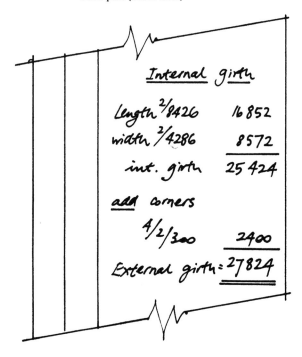

Figure 3.5 External girth. Here the waste calculation gives the external girth of the room, based on the internal plan dimensions.

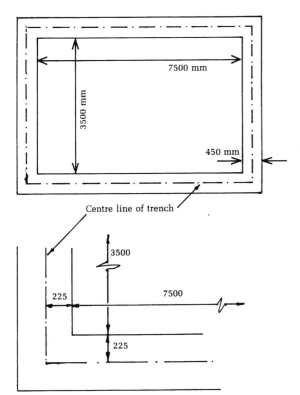

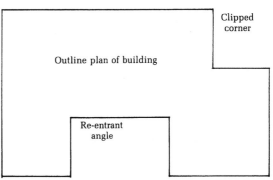

Figure 3.8 Clipped corners and re-entrant angles.

Figure 3.6 Foundation trench plan and corner detail.

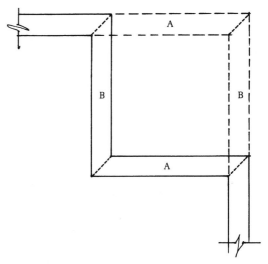

Figure 3.9 Clipped corner details.

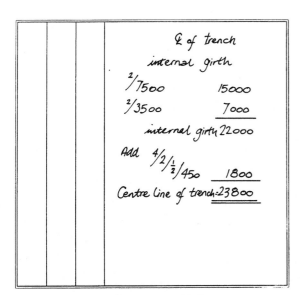

Figure 3.7 Waste calculation establishing the length of trench in Fig. 3.6.

building, provided that the corners are all right angles. From this example it can be seen that, so far as the girth is concerned, an internal angle compensates for an external angle.

Figure 3.10 shows an irregular plan shape with external angles marked 'x' and internal angles marked 'o'. A simple count of the external and internal angles will always reveal an excess of four external angles, so long as the angles are all regular and the outline of the building is fully enclosed. The dimensions used to establish the girth of this building need only relate to the overall lengths A and B. So far as a centre line calculation is concerned, the equation need only make an adjustment for an excess of four external angles.

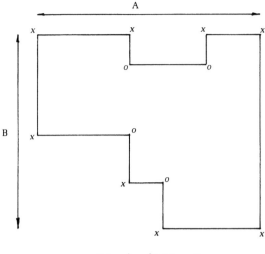

External angles (x) = 8
Internal angles (o) = 4

Excess of external angles = 4

Figure 3.10 Identification of internal and external corners.

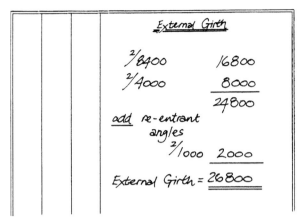

Figure 3.12 Waste calculation establishing perimeter including a re-entrant angle.

3.3.1 Re-entrant angles

In this case (figure 3.11) the plan shape is interrupted by a projection into the building (e.g. where a chimney stack projects into a room). This will have an effect on the perimeter length of the room. The calculation of the girth will include the length L above but will make no allowance for the depth of the projection D. These must be added to obtain the perimeter length. The waste calculation would be presented as in figure 3.12.

Care must be taken to study the drawings carefully when calculating perimeter lengths. In the majority of cases the perimeter calculation will be used as the basis for recording a number of dimensions which relate to it. Often the walls

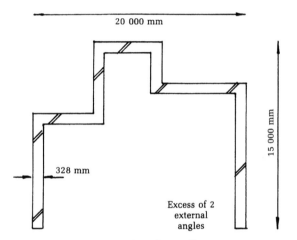

Figure 3.13 An external garden wall.

of an extension to a building or an external garden wall, such as in figure 3.13, will not form a complete enclosure and in these cases any adjustment for a centre line must take into account the number of external and internal angles.

In figure 3.13 the walling shows four external angles and two internal angles. Once the girth is established the adjustment for the centre line of the wall must allow for an excess of only two external angles (figure 3.14).

3.3.2 Irregular areas

By carefully dividing large irregular shaped areas it is possible to establish a number of smaller geometric forms. These can then be recorded as a series of triangles, rectangles and squares (figure 3.15).

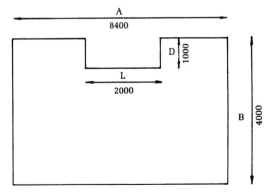

Figure 3.11 Re-entrant angle details.

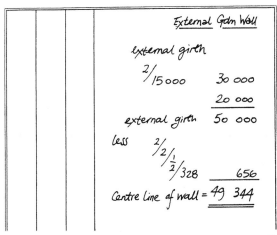

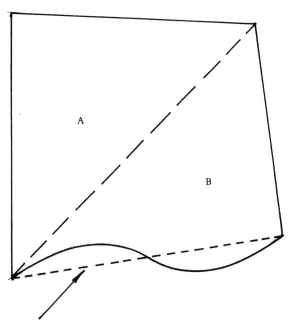

Give and take or compensation line

Figure 3.16 Compensating for an irregular boundary.

Figure 3.14 Establishing the length of the garden wall in Fig. 3.13.

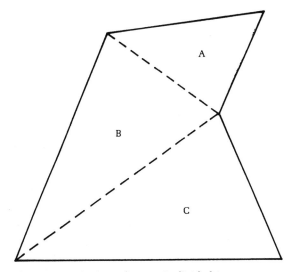

Figure 3.15 An irregular area is divided into a number of triangles.

evenly spaced offsets are available, the area may be divided into an even number of equally spaced strips and Simpson's Rule applied.

The quickest and most accurate approach is to use a digitiser. This allows the measurer to enter the appropriate scale and simply run the digitiser pen around the outline of the irregular area. The result is almost simultaneously computed and presented on the screen.

3.3.3 Triangles and circles

Refer to 'Geometric forms', p. 11 for examples of mensuration techniques associated with the measurement of triangles and circles.

3.3.4 Parallelograms and trapeziums

Where the two opposite sides of a four-sided shape are both parallel (parallelogram) the area can be recorded by multiplying the perpendicular height (H) by the length (L).

Where only two sides of a four-sided shape are parallel (trapezium) the perpendicular height between the parallel sides multiplied by the mean length of the parallel sides will give the area. The formulae for each of these, together with a number of other geometric forms, are given at the end of this book.

This approach works well when the sides of an irregular shaped area are straight. On occasions an irregular boundary line will require an adaptation of this approach. In order to approximate the area of the irregular plan shape shown in figure 3.16, the measurer will need to draw a straight line along the curved boundary to balance the area inside and outside this line. Having established this 'give and take line', the irregular area can be recorded using the same approach as before, as a series of triangles, rectangles and squares.

This is a rather haphazard approach which can be somewhat inaccurate. For a more accurate result and where

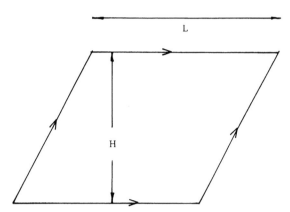

Figure 3.17   Area of a parallelogram.

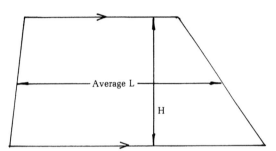

Figure 3.18   Area of a trapezium.

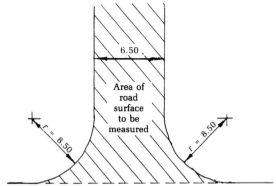

Figure 3.19   Outline plan of a bellmouth.

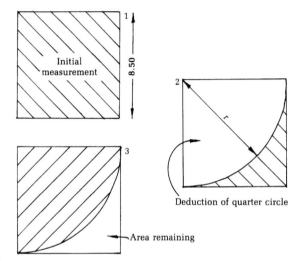

Figure 3.20   The three stages in the measurement of the curved area of a bellmouth.

3.3.5 Bellmouths

The term 'bellmouth' is used to describe the shape formed at a 'T' junction in a road. In the example shown in figure 3.19, in the first instance the road surface will be measured through the bellmouth at 6.50 metres width. This leaves two irregular curved areas at each side of the 'T' junction. By initially recording two squares, each of 8.50 metre side, and then deducting a quarter circle of 8.50 metre radius (again on each side) the correct area will remain (figure 3.20). In accordance with the basic principles of measurement, it is important to overmeasure in the first place and make adjustments after.

There are a number of other mensuration techniques which are used by measurers when taking-off quantities. Many of these are specific to certain types of work, e.g. groundworks, roof structures and roof coverings. Rather than record these here they have been included as a part of the introduction to their respective chapters.

4 Document production

4.1 Traditional BQ production

The principal purpose of measurement is to provide a Bill of Quantities (BQ) which can be used as the basis for the preparation of tenders for construction work. In due course the priced BQ will become one of the contract documents and legally bind the contractor to perform and the client to pay for the construction. In this form it provides a basis for the valuation of varied work and the preparation of interim or stage payments. It is also likely to be used by the contractor in the administration and organisation of construction operations.

Measurement can therefore be described as the starting point from which construction costs are established. In order for these measured items to accommodate the costing process they must be framed in a way that is meaningful to the sequence and division of operations on site. Costing a document comprising sporadic and unstructured measured items would be impossible.

Tendering arrangements may vary, but in the normal course of events the contracted works will be performed as a number of subcontract operations and it is important that the costing documentation replicates this pattern and distribution of work. There are a few anomalies but by and large the divisions of subcontracting operations are represented in this fashion by the Work Sections of SMM7. Where the other documents are prepared in accordance with the Common Arrangement of Work Sections (CAWS) these can be cross-referenced to each other, avoiding unnecessary repetition and providing a coherent structure for all tender documents.

Prior to the wide scale availability and adoption of computer technology the preparation of the BQ, by necessity, involved a number of tedious and repetitive operations culminating in the preparation of a handwritten draft. Once typed, and before being copied, the master would be read-over. The completed document would then be distributed with other tender documents to tendering contractors. There were a number of attempts made to rationalise this procedure. 'Billing direct' and 'cut and shuffle' are both examples of manual systems that were developed in order to avoid the need for abstracting.

Technology is available today which eliminates the intermediate stages traditionally carried out as part of the billing process. A library of standard descriptions held in a database can be matched with quantities digitised directly from the drawings. The squaring, transferring, merging and sequencing of items is carried out automatically. Not only does this significantly reduce the amount of time needed for BQ preparation, it also allows the contents of the Bill to be presented in a number of different forms. For example, when tendering, the Bill is likely to be prepared in Work Section order; when carrying out a valuation, the same measured items might be presented in elemental order.

With the development of CAD systems, BQ production software and standard libraries of specification clauses (NBS), all tender documentation can be generated electronically. At the same time many contractors maintain databases of production costs, and it would seem only a matter of time before all of these data can be exchanged electronically or in disk form. For practical and legal reasons there will continue to be a need for the production of a hard copy of working drawings and other tender documents.

4.2 The structure of the Bill of Quantities

Bills of Quantities should 'fully describe and accurately represent the quantity and quality of the works to be carried out'. So long as they achieve this there would appear to be no limit to their eventual form. Whilst there is every need to adopt a flexible approach to Bill preparation, it must produce a document with a recognisable structure which is consistently interpreted by its users. Most Bills of

13 Beams (nr) 14 Beam casings (nr) 15 Columns (nr) 16 Column casings (nr)	1 Attached to slabs 2 Attached to walls 3 Isolated	1 Regular shaped, shape stated 2 Irregular shaped, dimensioned diagram	m² m	1 Left in 2 Permanent 3 Curved, radii stated 4 Height to soffit ≤ 1.50 m 5 and thereafter in 1.50 m stages	M10 Passings of subsidiary beams or other projections are not deducted from areas of formwork but such intersections are deemed to constitute the commen...	D9 Where a downstand beam is formed by temporary formwork but the slab is supported by permanent formwork, the downstand beam regarded be...	C3 Formwork to beams, columns and casings is deemed to include ends	
17 Recesses (nr) 18 Nibs (nr) 19 Rebates (nr)	1 Plain rectangular, size stated 2 Dimensioned diagram		m	1 Extra over the formwork in which they occur 2 Left in 3 Permanent 4 Curved, radii stated	M13 The extra over superficially measured formwork in which recesses, nibs or rebates occur is stated		C4 Formwork to recesses is deemed to include ends	
20 Extra over a basic finish for formed finishes	1 Slabs 2 Walls 3 Beams 4 Columns 5 Others, stated		m²			D11 Formed finishes are those where a finish other than a basic finish is required		S3 Details of formed finishes

E20	FORMWORK FOR IN SITU CONCRETE (continued)		E20
	Formwork; basic finish E20/710		
	Soffits of slabs		
A	not exceeding 200 mm; thick; horizontal; propping height 3.00-4.50 m	186 m²	
	Beams attached to slabs		
B	plain rectangular; 300 x 450 mm; height to soffit 3.0-4.5 m (8 nr)	27 m²	
	Extra over basic finish for plain smooth finish E20/720		
C	slabs	162 m²	
D	beams	22 m²	

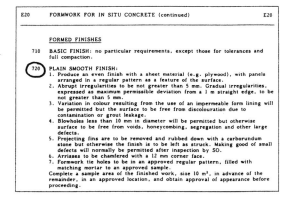

Figure 4.1 Use of SMM7 as a basic phraseology, including cross-reference to the specification. (Reproduced courtesy of the Royal Institution of Chartered Surveyors.)

Quantities are made up of a number of separate parts and may include separate sections for Preliminaries, Preambles, Measured Work, Work by Nominated Subcontractors/ Suppliers and Provisional Sums. Whilst SMM7 identifies these last two items as part of the Preliminaries, it is often found convenient for the preparation of valuations and final accounts if they are grouped together in their own separate section.

4.3 Bill preparation

In order to achieve consistent interpretation by all those who use the BQ a standard approach to billing will be necessary. Whilst the format and style of presentation may vary from one office to the next, the general principles remain the same. Standard ruled A4 size paper (known as single bill paper) is used for the presentation of billed work. The wide scale adoption of laser printers, word processing software and desktop publishing has enabled QS practices to develop variants to this standard BQ page. Whilst this has facilitated a more flexible interpretation in both style and format, it should not detract from the provision of consistent interpretation by all those who use the BQ.

The finished document should be typed on one side of the page and each Work Section be commenced on a new sheet. The extreme left-hand column (binding margin) is used to provide a system of indexing. This is necessary to allow all those who use the bill to locate and refer to specific bill items and is normally achieved by entering consecutive letters of the alphabet parallel with the first line of the description. When used in combination with the BQ page number this provides a unique reference for each bill item. Each page of the bill should give the job name and reference; the cash columns on the extreme right should be labelled with the appropriate Work Section and each cash column titled with pounds and pence.

The start of each Bill should give the bill number and the appropriate Work Section heading. Upper and lower case letters, bold face, underlining and indenting can be used to provide a structure and sequence to the descriptions. Assuming the adoption of a traditional BQ, the sequence of bill items will follow SMM7 Work Section order. Descriptions should be written in full to avoid the possibility of misinterpretation.

F MASONRY				£	p
	F10 BRICK/BLOCK WALLING				
	Facing brickwork above dpc F10/110				
	Walls				
A	Half brick thick; facework one side	253	m^2		
B	Half brick thick; curved on plan 1350 mm radius; entirely of headers; facework one side	36	m^2		
	Plain band				
C	75 mm wide; sunk 25 mm from face of wall; horizontal	32	m		
	Plinth capping				
D	Half brick wide; flush; horizontal; entirely of stretchers	12	m		
E	Half brick wide; flush; horizontal; entirely of headers	4	m		
	Facing brickwork below dpc F10/110				
	Walls				
F	Half brick thick; facework one side	24	m^2		
G	Half brick thick; curved on plan 1350 mm radius; facework one side	1	m^2		
	Facing blockwork F10/210				
	Walls				
H	100 mm thick; facework one side	438	m^2		
J	100 mm thick; curved on plan 1150 mm radius; facework one side	33	m^2		
K	100 mm thick; curved on plan 1150 mm radius; entirely of half blocks; facework one side	36	m^2		
L	Extra for special 90 deg dog-leg blocks	120	m		
M	Extra for special 190 mm deep lintol blocks (concrete and reinforcement measured separately)	99	m		
			TO COLLECTION		

3/16

Figure 4.2   Standard rulings of stationery for quantity surveyors (detailed in BS3327).

4.3.1 Units of billing

As part of the abstracting process the quantities will have been rounded to the nearest whole unit. Fractions of less than half a unit will have been disregarded, whilst those over a half will be taken as a whole unit. In some circumstances this may result in the elimination of a measured item (i.e. where the unit is less than 0.50), in which case the item will be given as a whole unit (SMM7 General Rule 3.3). Exceptionally quantities measured in tonnes should be given to two places of decimals. When entered in the unit column the following symbols are used to represent these units:

m = metre
m^2 = square metres
m^3 = cubic metres
nr = number
kg = kilogram
t   = tonne
h   = hour

4.3.2 Collections and summaries

At the foot of each page the cash columns should be ruled off and the words 'To Collection' entered. At the end of each Work Section a Collection is set up to provide a framework for the estimator to total the costs for the Work Section.

At the end of the measured Work Sections, a summary of the totals from all of the Work Sections is prepared. The total cost of the measured work will in turn be transferred to a General or Grand Summary which collects the costs from the other bill sections and provides the total to be transferred to the form of tender.

4.3.3 Sections of the BQ

The component parts of the BQ may include all or some of the following:

Preliminaries
Preambles
Measured Work
Nominated Subcontractors and Suppliers
Contingencies

These may be presented as separate sections or in some cases combined to provide conglomerate sections depending on personal preference or the particular requirements of the project. Where the structure of SMM7 is adopted, work by Nominated Subcontractors/Suppliers and any Contingencies will appear in the Preliminaries (SMM7 A51−55). Each part of the BQ would be represented by a section number (e.g. Section 1 Preliminaries, Section 2 Preambles,

Section 3 Measured Work Sections, etc.). Page numbers relate to each Section; thus, if there were 24 pages of Preliminaries these would be identified at the foot of each page by the Section number followed by the page number (1/1 to 1/24). Each description is identified by a reference, either alphabetic or numeric, which when used in combination with the page number provides a unique location reference.

Preliminaries The Preliminaries/General Conditions Section of the BQ provides the project particulars and identifies the works, the project and the parties to the contract. In addition they include provision for the costing of items that do not form a part of the permanent works but are nonetheless necessary for the successful completion of the project. Items such as insurances, temporary buildings, scaffolding, protection and the provision and maintenance of plant all fall into this category.

Preambles The Preambles define the quality of materials and the standard of workmanship for the project. They may be included in the BQ as a separate conglomerate section or broken down and presented with the Work Section to which they relate. Whilst they will inevitably affect the price of the Measured Work Section, it probably makes for better presentation if they are included in the finished document as a single separate section. Where the architect has prepared drawing specification references using the Common Arrangement of Work Sections (CAWS), the measured work items and drawing references will automatically cross-reference to the Preambles. The National Builders Specification (NBS) has been prepared to accord with CAWS and provides a valuable source of information in the preparation of both Specification and Preambles.

Measured Work Section In most cases this will be the largest section of the BQ and will include measured items presented in the sequence of SMM7 (see figure 1.6). It is most unlikely that the completed BQ will include each and every Work Section. The layout, page format and presentation of the Measured Work Section is given elsewhere in this chapter.

4.3.4 Prime Cost and Provisional Sums

It is not uncommon for the tender documentation to be prepared without the full details of the design of a building being available. Where the full nature and extent of the work is unknown this is unavoidable. In other cases a decision on the supply of materials or the services of a specialist subcontractor will only be made once contracts have been signed and work on site has commenced. In order that the

			GROUNDWORKS
			£	p

D   GROUNDWORKS

COLLECTION

Page 3/4
       3/5
       3/6
       3/7
       3/8
       3/9
       3/10
       3/11

TO SUMMARY   £

3/12

Figure 4.3   A collection page at the end of a Work Section.

	£	p

SUMMARY			
	Page		
D GROUNDWORKS	3/12		
F MASONRY WORK	3/20		
G STRUCTURAL/CARCASSING/METAL/ TIMBER	3/32		
H CLADDING/COVERING	3/46		
L WINDOWS/DOORS/STAIRCASES	3/58		
M SURFACE FINISHES	3/68		
P BUILDING FABRIC SUNDRIES	3/75		
Q PAVING/PLANTING/FENCING	3/82		
R DISPOSAL SYSTEMS	3/104		
T MECHANICAL HEATING	3/122		
V ELECTRICAL POWER/LIGHTING	3/146		

TOTAL CARRIED TO GENERAL SUMMARY £

3/150

Figure 4.4 The summary page of all Work Sections.

				£	p
GENERAL SUMMARY					
	Page				
SECTION 1 – PRELIMINARIES	1/24				
SECTION 2 – PREAMBLES	2/48				
SECTION 3 – MEASURED WORK	3/150				
SECTION 4 – NOM S/C AND SUPPLIERS	4/12				
SECTION 5 – CONTINGENCY	5/3				
TOTAL CARRIED TO FORM OF TENDER		£			
GS/1					

Figure 4.5   A General or Grand Summary.

costs of these items are included in the tender documentation, sums of money are allocated in the BQ by way of Provisional or Prime Cost Sums. An adjustment will be made in the Final Account by setting actual costs against the inclusion for Prime Cost or Provisional Sums.

Prime Cost Sums These include work executed by specialist subcontractors or material suppliers nominated or selected by the architect or client. The contractor is entitled to recover any costs associated with the nomination, such as attendances and loss of profit, and the BQ is prepared to enable tenderers to make an inclusion for these. In order to complete the work, nominated subcontractors may need the use of temporary lighting, storage facilities and scaffolding. (A full list is given in SMM7 A42, A43 and A44.) These are termed Contractor's General Cost Items or General Attendance Items and will be provided as a matter of course by the main contractor for the execution of the works. Since these facilities are available anyway, the additional cost associated with nominated subcontractor use is minimal. However, they are technically the property of the contractor and the opportunity for these to be costed must be made at the tender stage. Where the nomination is for work of a specialist nature, or where it involves the provision of facilities which are not defined as general cost items, the BQ should make provision for the inclusion of Special Attendance items which are additional to the main contractor's provision (SMM7 A51.1.3).

It is possible for Prime Cost and Provisional Sums to appear in any one of three different locations in the completed BQ. They may be included with the Preliminaries, with the Measured Work Sections or in a separate section of their own. Wherever they appear, they would typically be presented as shown in figures 4.6 and 4.9.

Provisional Sums Where there is insufficient information available for proper measurement a Provisional Sum can be written into the BQ. An inclusion may also be made where there is an element of doubt over actual costs but where the extent and nature of the work will become apparent once work on site has commenced. Two categories of Provisional Work are identified in SMM7, defined and undefined.

Defined Provisional Sums (figure 4.8) are sums allocated for the performance of specific tasks. The nature, method and extent of the work would only become apparent once work on site had been commenced. Undefined Provisional Sums (figure 4.9) are more difficult to describe and would generally embrace sums of money written into the BQ for unforeseen circumstances by way of a contingency. The cost of any additional substructures, the provision of samples and the carrying out of tests may be included as an Undefined Provisional Sum.

A51 NOMINATED SUBCONTRACTORS		Preliminaries	
		£	p
	Nominated Sub Contractors		
a	Include the prime cost sum of £4,800 for electrical power/lighting installation in accordance with Specification ref V10.8 and Drawing ref: Elect V006	4,800	00
b	Add main contractor's profit %		

Figure 4.6 The presentation of a Prime Cost Sum for work to be carried out by a Nominated Subcontractor.

A53 WORK BY STATUTORY AUTHORITIES					Preliminaries	
					£	p
	Work by Statutory Authorities					
	Include the following provisional sums for work carried out by Public Undertakings and Statutory Authorities. The Contractor is invited to add for any profit and loss of discount. The amount included in the final account will be based on the actual charge made by the Statutory Authority.					
a	Southern Electricity				650	00
b	British Telecom				250	00
c	Southern Water				1,500	00

Figure 4.7   The presentation of a Prime Cost.

A54 PROVISIONAL WORK					Preliminaries	
					£	p
	Provisional Work					
	Include the following defined provisional sums					
	Supply only					
a	Kitchen wall and floor units				5,200	00
b	Sanitaryware				2,400	00
c	Ironmongery				800	00

Figure 4.8   The presentation of a Provisional Sum.

A54 PROVISIONAL WORK					Preliminaries	
					£	p
	Provisional Work					
	Include the following undefined provisional sums					
a	Contingencies				2,500	00
b	Testing, samples				200	00

Figure 4.9   Undefined Provisional Sums.

5 Groundworks

5.1 Introduction

This Work Section of SMM7 includes the measurement rules for a number of construction operations associated with structural or supporting work in the ground. These include piling, diaphragm walling and underpinning together with excavation and filling. Consideration is given here for the substructure operations associated with simple domestic construction including reduce level excavation, trench fill and strip foundations.

In the normal course of events the construction of foundations will require the excavation of trenches, the casting of concrete foundations and the laying of brickwork or blockwork to support a ground floor slab. In addition to the Groundworks Section of SMM7, the measurer must therefore be familiar with some of the measurement rules for both In-situ/Precast Concrete Work and Masonry Work Sections. Many surveyors choose to group these three separate Work Sections under a single common Bill of Quantity heading of Substructures. All work up to and including the damp proof course/membrane together with the structural ground floor may be included in the Substructure measurement.

In summary, the term 'Substructures' describes all structural work below ground level. It is likely to include the SMM7 Work Sections D (Groundwork), E (In-situ Concrete/Large Precast Concrete) and F (Masonry). The demarcation between sub- and superstructure should be taken as the damp proof course in structural walls. The structural floor slab would be measured with substructure work, whilst the screed and any floor finish would be included with Work Section M (Surface Finishes).

5.2 Excavation

When measuring groundwork (excavation) for simple domestic dwellings it will be necessary to book dimensions for the following items: excavating topsoil, excavating to reduce ground levels and excavating foundation trenches. In addition to each of these, it will be necessary to support the sides of trenches, compact and possibly blind the bottoms of trenches, remove any excavated materials from or about the site and dispose of any water that may be present during these operations. Each of the excavation items given above are summarised in table 5.1 together with their appropriate disposal item.

Before commencing measurement it will be necessary to study the drawings, specification and any borehole log details to identify the following:

- Existing ground levels
- Finished ground levels
- Finished floor levels
- Foundation depths and widths
- The ground water level and the date of its reading
- The existence of known underground services or obstructions.

The cost of excavation is dependent on a number of factors. To allow the estimator the opportunity to cost these it is important to identify the type of ground to be excavated,

Table 5.1 Excavation measurement items

Operation	Items to be measured	Unit of measurement	SMM clause
Excavate topsoil	1. Excavation 2. Disposal	m^2 m^3	D20.2.1.1 D20.8.3.1/2
Excavate to reduce levels	1. Excavation 2. Disposal	m^3 m^3	D20.2.2.1–4 D20.8.3.1/2
Excavate trenches	1. Excavation 2. Disposal	m^3 m^3	D20.2.5/6,1–4 D20.8.3.1/2

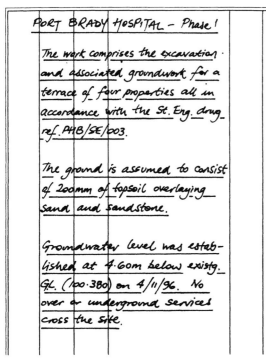

Figure 5.1 An example heading page of a take-off.

D20.P1 requires all of these data to be detailed as part of the completed Bill of Quantities. Initially this can be included as a heading in the take-off and this will later be transferred to provide a heading in the Substructure Work Section of the Bill of Quantities.

5.2.1 Surface excavation

The term 'surface excavation' refers to the excavation of the surface of the site and describes the excavation of topsoil and/or excavation to reduce levels. The surface of most 'green field' sites comprises a compactable layer of vegetable matter called 'topsoil'. Where space permits it can be carefully removed and stored on site for later reuse. In such cases, it is measured in square metres, stating the average topsoil depth in the description (SMM7 D20.2.1.1). Alternatively, if there is no topsoil, or if it is not required on site, it can be grouped with the reduce level excavation and measured in cubic metres, stating the maximum depth of dig in the description (SMM7 D20.2.2.1–4).

The plan dimensions for surface excavation must include the spread of the foundations beyond the face of the external walls. These may not be immediately available from the site plans, since the dimensions on the drawing provide only external wall dimensions. (As far as measurement is concerned, it is only necessary to measure the minimum excavation required for foundations; in practice, it is unlikely that a contractor will excavate within this minimum.) Given the width of external wall foundations, together with the thickness of the external walling, it is possible to calculate the projection of the foundation beyond

its stability and cohesion, whether the excavation will be affected by the presence of groundwater and the ease with which earthmoving plant can move on and about the site. Where a site survey has been carried out this information will be recorded on the site survey drawings. SMM7

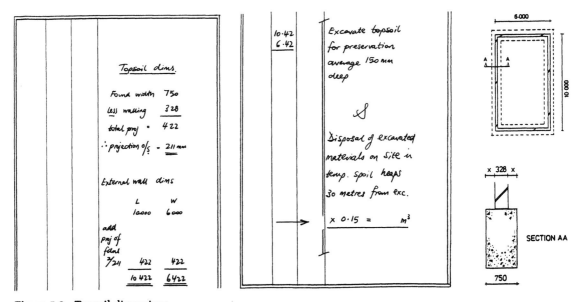

Figure 5.2 Topsoil dimensions.

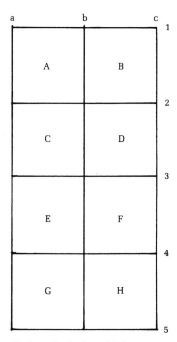

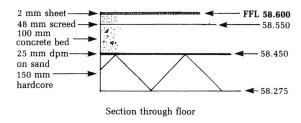

Section through floor

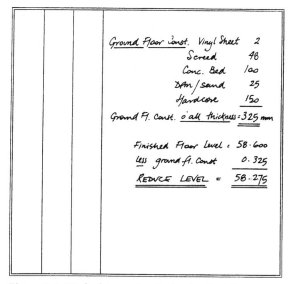

Ground Floor Const. Vinyl Sheet 2
                       Screed   48
                    Conc. Bed   100
                    DPM / sand   25
                    Hardcore   150
Ground Fl. Const. o'all thickness = 325 mm

Finished Floor Level = 58·600
less ground fl. Const   0·325
REDUCE LEVEL =   58·275

Figure 5.4 Worked example of a level calculation.

Figure 5.3 Grid method of establishing average ground level. Depending on its position in the grid, each point will occur once, twice or four times in the averaging calculation. For example, point **b1** occurs twice, once in the calculation of square A and once in the calculation for square B. Point **b4** occurs four times, once for each square E, F, G and H. The points at the four corners are each used only once. The level of each point is listed in turn and then multiplied by the number of times it should occur in the overall calculation (i.e. its weighting). If this total is divided by the total number of grid points used (four for each square), the result is the average existing ground level.

the external wall. Working on the reasonable assumption that the walling is central to the foundation, then the projection on either side of the foundation will be the same.

Few sites are level and the majority will require some excavation to provide a horizontal surface from which construction operations can commence. The removal of material in this way is measured in cubic metres by booking the plan dimensions, together with the average depth of the excavation (SMM7 D20.2.2.1−4). Depending on the substructure construction, it may be necessary to measure working space in addition to the reduce level excavation (SMM7 D20.6.1.*). As stated previously, the plan dimensions for reduce level excavation should include the projection of foundations beyond the face of external walls. The depth will obviously vary and will need to be recorded in the dimension column as an average. Where a grid of levels is available, this can be carried out as shown in figure 5.3. Note that where a cut and fill line occurs (see section 5.3),

average ground levels are calculated separately on either side of this line.

A level is recorded at each intersection and corner of the grid. If each individual grid square is averaged and the total of these subsequently averaged, the overall average ground level can be found. This can be an extended and cumbersome waste calculation, but it is important to show this approach in order to demonstrate the principle adopted in the more efficient technique of 'Weighting'.

5.2.2 Establishing reduce level

The next stage is to identify the ground level on the underside of the floor slab. This is the level that will form the base for the subsequent floor construction. From a section through the proposed ground floor slab construction the total thickness of the ground floor construction can be found. Once this is known, it is deducted from the finished floor level to provide the reduce level required. Having established the existing ground level and the required level, the average reduce level excavation depth is found by deducting one from the other.

The following example (figure 5.5) shows the waste

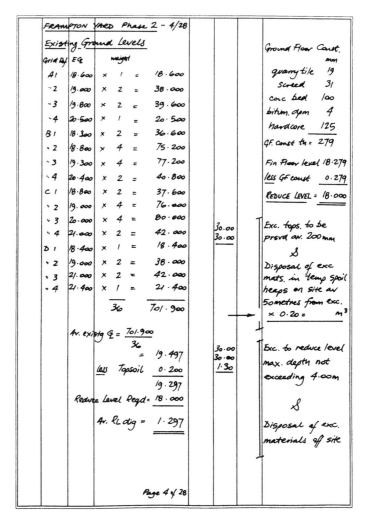

Figure 5.5 Waste calculations and booked dimensions for a topsoil and reduce level excavation.

calculations and subsequent booked dimensions for the removal of topsoil and reduce level dig. The extreme dimensions of this oversite excavation are assumed to be 30 m × 30 m and the reduce level required is given at 18.000. (This exercise includes a waste calculation that would normally precede the booked dimension and is presented in this fashion simply because of the limitations of space.)

The description for reduce level excavation should give the maximum depth range, rather than the average depth required over the entire site.

5.2.3 Disposal of excavated materials

When any excavation work is carried out it will tend to increase its bulk since air pockets are created in what was previously undisturbed ground. No adjustment is made to the quantities for this increase in the volume of material caused by 'bulking' (see SMM7 General Rule Clause 3.1, together with Section D20 Measurement Rules M3 and M13). The increase in the volume of excavated material will vary, depending on the type of material excavated. An appropriate allowance made by the estimator will accommodate this apparent increase in volume when the substructure work is costed.

Removing excavated material from site often costs more than the initial excavation. For this reason it is always necessary to measure disposal of excavated materials as a *separate item* (i.e. own description and quantities) (see table 5.1). Clauses D20.8.3.1 and 2 provide the alternatives for disposal of excavated materials.

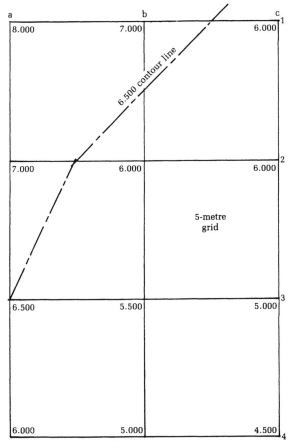

In this example a cut and fill line of 6.500 can be readily identified and positioned between the given grid points. Consider grid line **1**: if the slope of the ground is assumed to be consistent, then the contour line for 6.500 will fall half-way between the grid points 6.000 and 7.000. The same exercise is carried out on grid lines **b** and **2**, whilst the cut and fill line coincides with the grid point **3a**. These points are then joined together to give a contour line representing the cut and fill line. The site levels that are higher than this line will be excavated, whilst those below will be filled. Each will need a waste calculation to determine the average dig depth and the average fill depth. Once these are known, the geometric forms representing the plan shape of the cut and fill can be recorded in the dimension column, followed by their respective average depths, thereby giving the volume of material involved.

Figure 5.6 Cut and fill line.

When the Bill of Quantities is prepared, all excavation items are classified in accordance with those given in SMM7 D20.2. Each of these will in turn generate its own disposal item which will be either on site (D20.8.3.2) or off site (D20.8.3.1). No reference is made to where this material came from (trench excavation, basement excavation or reduce level excavation); it is all simply described as 'excavated material for disposal'. However, it will be necessary to identify where and how the material is to be transported, e.g. 'temporary spoil heaps maximum 50 metres from the excavation' (D20.8.3.*.1 or 2).

5.3 Cut and fill

As an alternative to excavating the surface of a site to obtain a level surface, a combination of excavation and filling can achieve some savings in the cost of both excavation and disposal. This presumes that the excavated material is suitable for the purposes of filling and that adequate

precaution is taken to ensure that this material is thoroughly compacted. Often, excavated material will be unsuitable as fill and any savings will be offset by the provision of imported fill material.

When a site is to be cut and filled, the measurer should plot on the grid of levels a 'cut and fill' line. This line represents the point where there is no excavation or filling. On one side of the line the site is excavated, on the other it is filled. Where the excavated material can be used as oversite filling, the cost of disposal can be minimised.

Few cut and fill excavations are as straightforward as the example shown in figures 5.6 and 5.7. Frequently, the positioning of a cut and fill line will require other considerations, such as the removal of topsoil and excavating and filling in depth ranges. It is very unlikely that a cut and fill line will fall consistently at a mid-point between two whole number grid points.

Where the positioning of a cut and fill contour line is less obvious, the following approach can be adopted (refer

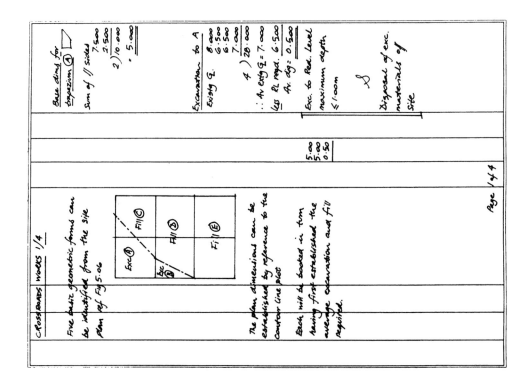

CROSSROADS WORKS 2/4

Excavation to B

Exstg G.L   7.000
            6.500
            6.500
         3) 20.000

∴ Av exstg G.L = 6.666
Less R.L. reqd   6.500
Av dig =   0.166

But Av dig over area
Ⓐ + Ⓑ as one site op.

∴   0.500
    0.166
 2) 0.666 = 333mm

∴ Same depth class
i.e. ≤1.00m deep

Exc. to Red Level
maximum depth
≤1.00m

∫

Disposal abd

½/ 5.00
    2.50
    0.17

CROSSROADS WORKS 1/4

Five basic geometric forms can
be identified from the site
plan of Fig 5.06

The plan dimensions can be
established by reference to the
contour line plot

Each will be treated in turn
having first established the
average excavation and fill
required.

Base dims for
trapezium Ⓐ

Sum of // sides   7.500
                  2.500
               2) 10.000
                = 5.000

Excavation to A

Exstg G.L   8.000
            6.500
            6.500
            7.000
         4) 28.000

∴ Av exstg G.L = 7.000
Less R.L reqd   6.500
Av dig =   0.500

Exc. to Red Level
maximum depth
≤1.00m

∫

Disposal of exc.
materials of
site

5.00
5.00
0.50

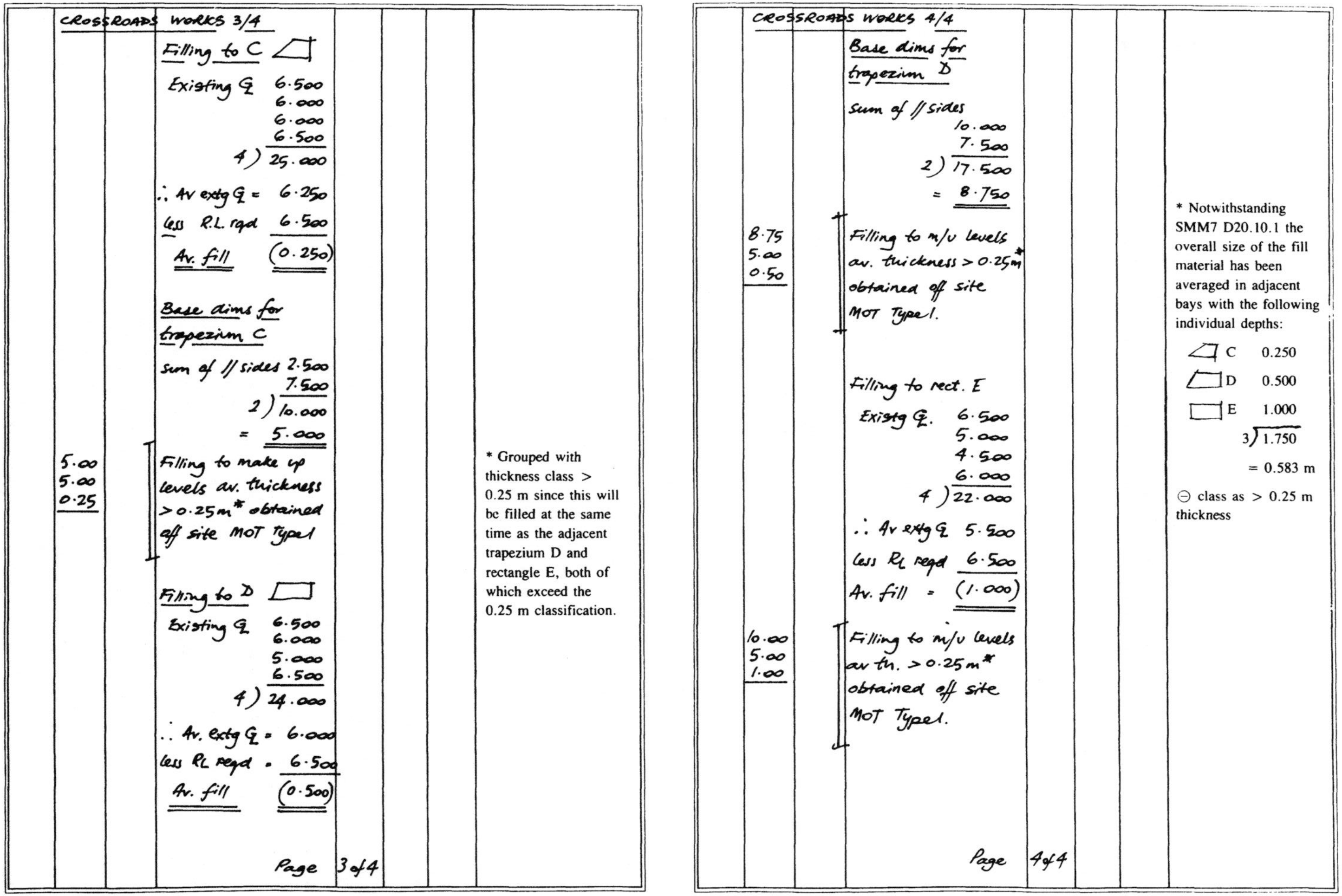

CROSSROADS WORKS 4/4

Base dims for trapezium D

Sum of // sides 10.000
                  7.500
              2)17.500
              = 8.750

8.75	Filling to m/u levels > 0.25m
5.00	av. thickness obtained off site
0.50	MOT Type I.

Filling to rect. E

Existing GL   6.500
              5.000
              4.500
              6.000
          4)22.000
          = 5.500

∴ Av extg GL = 5.500
less RL reqd   6.500
Av. fill =   (1.000)

10.00	Filling to m/u levels
5.00	av. th. >0.25m*
1.00	obtained off site
	MOT Type I.

* Notwithstanding SMM7 D20.10.1 the overall size of the fill material has been averaged in adjacent bays with the following individual depths:

C 0.250
D 0.500
E 1.000
3)1.750
= 0.583 m

① class as > 0.25 m thickness

Page 4/4

CROSSROADS WORKS 3/4

Filling to C

Existing GL   6.500
              6.000
              6.000
              6.500
          4)25.000
          = 6.250

∴ Av extg GL = 6.250
less RL reqd   6.500
Av. fill =   (0.250)

| 5.00 |
| 5.00 |
| 0.25 |

Base dims for trapezium C

Sum of // sides 2.500
                 7.500
             2)10.000
             = 5.000

Filling to make up levels av. thickness >0.25m* obtained off site MOT Type I.

Filling to D

Existing GL   6.500
              6.000
              5.000
              6.500
          4)24.000

∴ Av extg GL = 6.000
less RL reqd   6.500
Av. fill =   (0.500)

* Grouped with thickness class > 0.25 m since this will be filled at the same time as the adjacent trapezium D and rectangle E, both of which exceed the 0.25 m classification.

Page 3/4

Figure 5.7 Example take-off for a cut and fill excavation.

to figure 5.8):

(1) Decide where the cut and fill contour line cuts through the grid.
(2) Calculate the fall between these two grid points.
(3) Calculate the fall between the point required (R.L.) and the lower grid point.
(4) Divide the result of (3) by (2) and multiply the result by the size of the grid.
(5) Scale or plot the answer from the lower grid point.
(6) Repeat (1) to (5) for each grid line that is cut by the contour line.

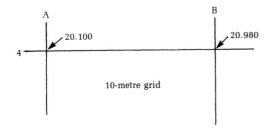

Plot grid point 20.850

1. Determine grid level required: 20.850
2. Calculate fall between two grid points:

    (B4) 20.980 − (A4) 20.100 = 0.880

3. Difference between point required and lower grid point:
    20.850 − 20.100 = 0.750

4. $\dfrac{\text{Point required}}{\text{Total fall}} \times \text{Grid size}$

    $\dfrac{0.750}{0.880} \times 10.000 = 8.520$ m

5. Scale answer (8.52 m) from the lower grid point (A4)

Figure 5.8 A more complicated cut and fill calculation.

5.4 Trench fill and strip foundations

Once the surface excavation has been completed, the measurement of foundation trenches, pits, etc. can commence. The starting level for this excavation is determined by the reduced level. Where this is more than 250 mm below the existing ground level, this must be stated in the description (SMM7 D20.2.*.*.1).

5.4.1 Trench fill

The simplest, and some argue cheapest, form of substructure is a trench fill foundation (figure 5.9). It is also a useful starting point for introducing the measurement techniques normally associated with foundation work. Table 5.2 provides a take-off list together with an indication of the unit of measurement, SMM7 reference and appropriate base dimensions for a trench fill foundation, though few

Table 5.2 Basic take-off list for trench fill foundations

Item	Unit	SMM7 ref.	Base dimensions
1. Excavate foundation trench	m³	D20.2.6	Length (CL of founds) × width × depth
2. Disposal excavated mats	m³	D20.8.3	Length (CL of founds) × width × depth
3. Earthwork support	m²	D20.8.7	(Ext. trench line × trench height) + (Int. trench line × trench height)
4. Surface treatments, compacting	m²	D20.13.2	Length (CL) × width
5. Concrete in foundation trench	m³	E10.1	Length (CL) × width × height
6. Masonry work to dpc (cavity work — each skin measured separately on own centre line)	m²	F10.1	Length (CL) × height
7. Topsoil backfill	m³	D20.9	Length (CL of fill) × width × height
8. Cavity fill	m³	E10.8	Length (CL of cavity) × width × height
9. Damp proof course (cavity work — each dpc measured separately)	m²	F30.2	Length (CL of wall) × width
10. Disposal of surface water	Item	D20.8.1	

foundations are as straightforward as this list would suggest.

Assuming the site has already been stripped of topsoil and that any reduce level excavation has been completed, the approach of table 5.2 can be adopted. (Note that trenches under 300 mm wide should be kept separate from those exceeding 300 mm wide.) If the construction work includes internal wall foundations, it would be normal to measure each in turn, taking care to note the width of internal foundations, since these are often narrower than their external wall equivalent.

In practice, any or all of the situations given in table 5.3 may be included, depending on the circumstances.

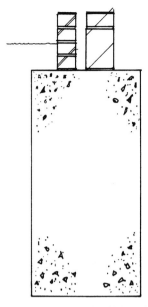

Figure 5.9 Trench fill foundation.

5.4.2 Strip foundations

Strip foundations differ from trench fill foundations in that a band of concrete is placed in the bottom of the trench which, in conjunction with masonry work and backfilling material, forms the basis of support for the superstructure. The approach to measurement is, in most respects, identical to trench fill foundations and differs only in the measurement of backfill material. Figure 5.10 shows backfill to

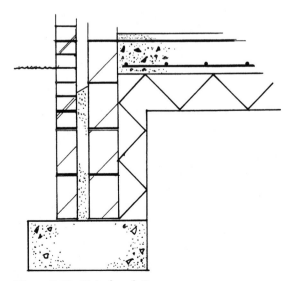

Figure 5.10 Strip foundation.

Table 5.3 Trench fill foundation measurement items

Situation	Unit	SMM7 ref.	Treatment
1. Excavating below ground water level*	m³	D20.3	Measure *extra-over* work previously booked
2. Excavating next to existing services	m	D20.3	Measure *extra-over* work previously booked
3. Excavating around existing services	m	D20.3	Measure *extra-over* work previously booked
4. Breaking out existing hard materials met in excavation	m³	D20.4	Measure *extra-over* work previously booked
5. Breaking out existing hard materials met in surface excavation	m²	D20.5	Measure *extra-over* work previously booked
6. Where face of excavation is less than 600 mm from face of formwork, rendering, tanking or protective walling	m²	D20.6	Measure girth of formwork etc. × depth of excavation (see Measurement Rule M8 of section D)

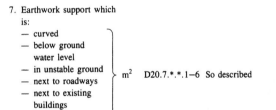

7. Earthwork support which is:
— curved
— below ground water level
— in unstable ground
— next to roadways
— next to existing buildings
— left in
} m² D20.7.*.*.1–6 So described

* See implication of this on measurement of earthwork support (measurement notes M10 and M11 of section D) together with disposal of ground water, ref. D20.8.2 (measurement note M5 of section D).

a strip foundation. This should be measured as filling to excavations (SMM7 D20.9.*.*.*) in cubic metres. The material used and the method of backfilling will depend on the substructure construction, but it is worth remembering that backfilling with excavated material below a solid ground floor should be avoided, and that backfilling with satisfactorily compacted hardcore would only be acceptable where the fill material is less than 600 mm thick or where the floor slab design includes reinforcement.

Bearing the above in mind, the following assumes an independent ground floor construction with floor loading passing to external and internal walls (figure 5.11). The following is traditionally recognised as the approach to be adopted when booking dimensions where backfill with

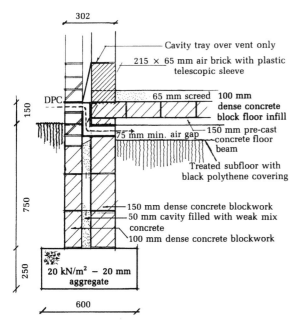

302

DPC

150

750

250

600

Cavity tray over vent only

215 × 65 mm air brick with plastic telescopic sleeve

65 mm screed   100 mm dense concrete block floor infill

75 mm min. air gap

150 mm pre-cast concrete floor beam

Treated subfloor with black polythene covering

150 mm dense concrete blockwork
50 mm cavity filled with weak mix concrete
100 mm dense concrete blockwork

20 kN/m² – 20 mm aggregate

Figure 5.11 Independent ground floor construction.

excavated materials is required; the notes and drawings are intended to help understand the principles used. The example in figure 5.11 assumes a trench centre line of 36.00 m; that backfill material is to be obtained from trench excavation and that any surplus excavated material will be removed from site. Other items that would normally be measured have been omitted.

The stages for establishing the quantity of filling and disposable excavated material are as follows.

Stage 1 Book the volume of trench excavation together with an equal volume of filling. The filling measurement is a theoretical paper one. The trench will not be completely backfilled on site until the structural walls and strip foundations are in place.

Stage 2 Concrete in the foundation trench is measured which will displace the fill material covered by the theoretical Stage 1 measurement. The now partly redundant fill material is no longer required and can be measured as disposal of excavated material.

Stage 3 Along the same lines as Stage 2, masonry work is measured in the foundation trench which will once again displace the theoretical fill material previously booked. Similarly, part of the fill material is no longer required and can also be measured as disposal of excavated material. However, the masonry work will be measured up to dpc level and in m². The adjustment of the filling materials

need only be made up to ground level but must be converted to m³.

5.5 Quantities

At this stage it is appropriate to establish the actual quantities of excavation, disposal and backfill. This can be carried out by preparing a mini abstract using the traditionally recognised procedure for booking dimensions where excavated material can be backfilled around foundations. As has already been mentioned there are some occasions when this technique is inappropriate. Where a dependent floor slab is adopted, current building practice is more likely to provide a detail similar to the one shown in figure 5.13.

In this example, imported hardcore must be used as backfill below a dependent slab, in combination with slab reinforcement. The backfill material used externally differs from the material used internally and is an irregular shape. In this situation it would be more appropriate to adopt the following approach:

(1) Excavate foundation trench.
(2) Dispose of all excavated materials off site.
(3) Find centre line of hardcore backfill and book volume (see note below).
(4) Find centre line of excavated material backfill and book volume.
(5) *Deduct* the same volume as in (4) from material previously disposed off site (this should be added on to item (4)).
(6) *Add* the same volume as in (4) but this time describe as disposed of on site in temporary spoil heaps.

The awkward sectional shape (parallelogram) of hardcore backfill must be booked as a volume and can be broken down into two simpler shapes (a rectangle and a triangle) with each booked as a volume on their individual centre lines.

5.6 Ground floors

To complete the work that would normally be included as part of substructure measurement we must now consider ground floors. For most domestic construction these can be classified as follows:

- Suspended timber (partially independent)
- Solid slab (dependent)
- Concrete beam and block infill (independent)

5.6.1 Suspended timber ground floors

Whilst these have been popular in the past, their use today is infrequent. As the title suggests, this method of ground

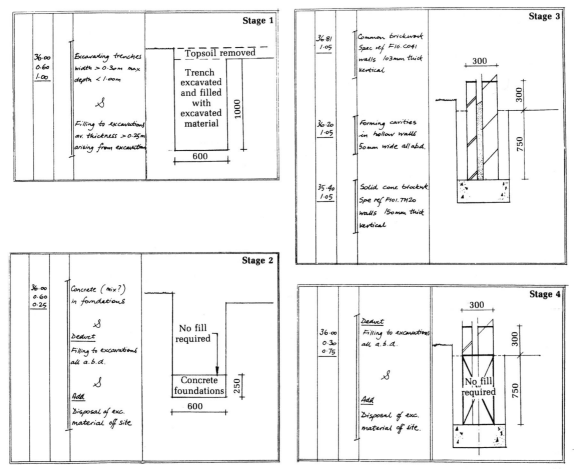

Figure 5.12   Establishing the quantity of filling and disposable excavated material.

floor construction is independent of the strata immediately below the floor. Ground floor loads are transmitted via external walls to the foundations or, where spans are excessive, by the inclusion of sleeper walls (see figure 5.14).

Where fill is incorporated below a ground floor which is more than 600 mm deep, it is advised that a suspended or independent floor construction is used (NHBC Practice Note 6). Measurement will usually include the items given in table 5.4.

5.6.2 Solid concrete ground floor slabs

This floor type bears directly onto the ground or filling layer immediately below. Its effectiveness depends on how well the filling layer has been compacted, as ground floor loads are transmitted directly through the fill material. The failure

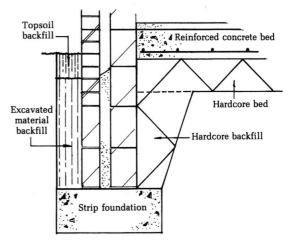

Figure 5.13   Strip foundation and backfill.

Table 5.4 Suspended timber ground floor measurement items

Item	Unit	SMM7 ref.	Base dimensions
1. Filling to make up levels	m³	D20.10.*.3	Plan length × plan width × depth
2. Surface treatments, compacting	m²	D20.13.2.2	Plan length × plan width
3. Concrete bed	m³	E10.4.*	Plan length × plan width × depth
4. Reinforcement fabric	m²	E30.4.1	Plan length (less cover) × plan width (less cover) (where sleeper walls bear on slab)
5. Sleeper walls	m²	F10.1.*.*	Length × height (honeycomb bond)
6. Damp proof course	m²	F30.2.*.*	Length × width
7. Air bricks	nr	F30.12.1	
8. Wall plate	m	G20.8	Length
9. Floor joist	m	G20.6	Length
10. Floor boarding	m/m²	K20.2	Plan length × plan width

Table 5.5 Solid concrete ground floor slabs measurement items

Item	Unit	SMM7 ref.	Base dimensions
1. Filling to make up levels (not to exceed 600 mm thick)	m³	D20.10.*.3	Plan length × plan width × depth
2. Surface treatment, compacting and blinding	m²	D20.13.2.2	Plan length × plan width
3. Concrete bed	m³	E10.4.*	Plan length × plan width × depth
4. Formwork (if necessary)	m	E20.2.*.*	Perimeter length
5. Fabric reinforcement	m²	E30.4.1	Plan length (less cover) × plan width (less cover)
6. Damp proof membrane:			
— flexible sheet	m²	J40.1	
— liquid applied	m²	J30.1	Length × width
— mastic asphalt	m²	J20.1	

Note: Floor screeds do not usually form a structural element of floor construction and would be measured as part of finishes (SMM7 M10).

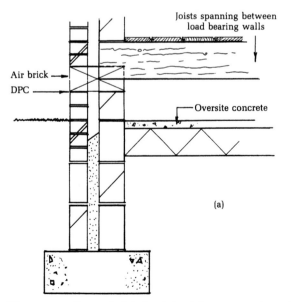

(a)

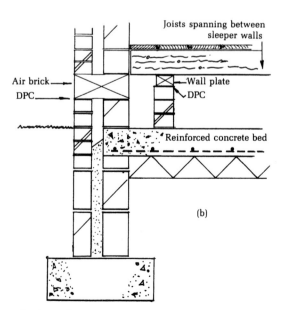

(b)

Figure 5.14 Suspended timber joisted floors: (a) independent; (b) dependent.

to correctly compact fill material has been the most frequent major defect in new houses where solid floor construction is used and is most apparent where floor slabs are carried off deep fill. The introduction of fabric reinforcement to the slabs together with slabs bearing on the inner skin of cavity walls will reduce the risk of slab settlement.

Measurement would usually include those items given in table 5.5.

5.6.3 Concrete beam and block infill floors

This is increasingly popular as a method of avoiding problems of settlement caused by solid slab floors. The system works on the same principle as a suspended timber floor by transmitting floor loads via structural floor beams to external and internal load-bearing walls. The beams are spaced at intervals that correspond with standard concrete block sizes, allowing blocks laid flat to complete the floor surface.

5.7 Worked take-off example

A worked example of a take-off for a concrete beam and block infill floor is shown on the following pages (take-off sheets: pp. 52−8; drawings: pp. 58−9).

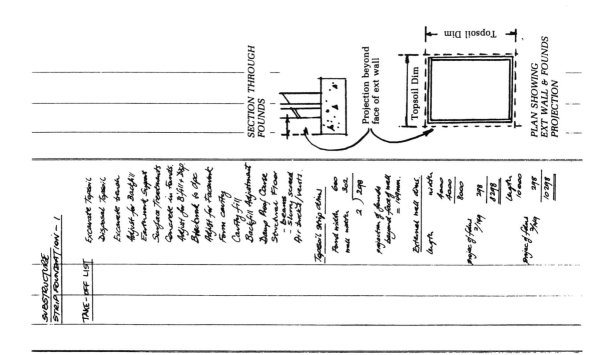

SUBSTRUCTURE / STRIP FOUNDATION-2

10.30 8.30	Excavating topsoil for preservation average 150mm deep
DDT 4.00 3.00	Disposal excavated material on site in temporary spot heaps 20m. from excavation
	× 0.15 = ___ m³

Centre Line of Trench.

External wall dims
2/10.000 20.000
2/8.000   16.000
          36.000
Ext gpth but = 36.000
Adjust for R.
Less 4/2/3/352 / 208
C.L. of trench = 34.792

Notes (right margin):

SMM7 D20.2.1.1 Excavation measured m².

SMM7 D20.8.3.1/2 Disposal of excavated material measured in m³.

Convert previously recorded area to a volume using constant dimension – in this case 150 mm depth of topsoil.

Centre line calculation based on ext wall dimensions assumes cavity wall is central to trench.

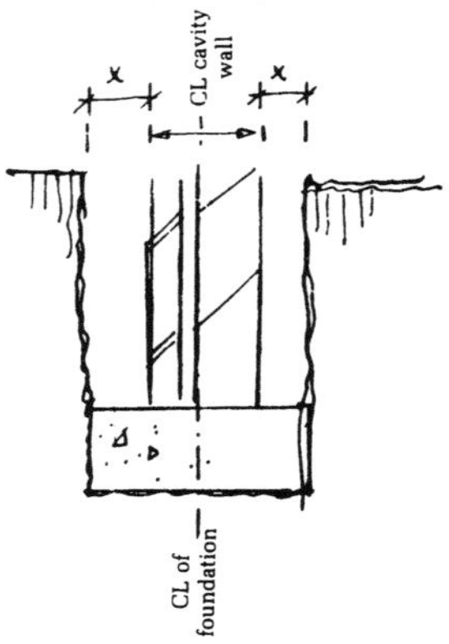

CL cavity wall / CL of foundation

SUBSTRUCTURE / STRIP FOUNDATION-1

TAKE-OFF LIST

Excavate Topsoil
Disposal Topsoil
Excavate trench
Adjust for Backfill
Earthwork Support
Surface Treatments
Concrete in Fonds.
Adjust for B.fill + Disp.
Blockwork to dpc
Adjust for Facework
Form cavity
Cavity Fill
Backfill Adjustment
Damp Proof Course
Structural Floor
 - beams
 - Slurry screed
Air bricks/vents.

Topsoil strip dims
Fond width   600
wall width   302
          2) 298
projection of founds beyond face of wall = 149mm.

External wall dims
length   width
2/4.000   298
 4.000   8.298
 8.000
              8.298

length
2/10.000   298
 10.000   10.298
              10.298

SECTION THROUGH FOUNDS

Projection beyond face of ext wall

Topsoil Dim

Topsoil Dim

PLAN SHOWING EXT WALL & FOUNDS PROJECTION

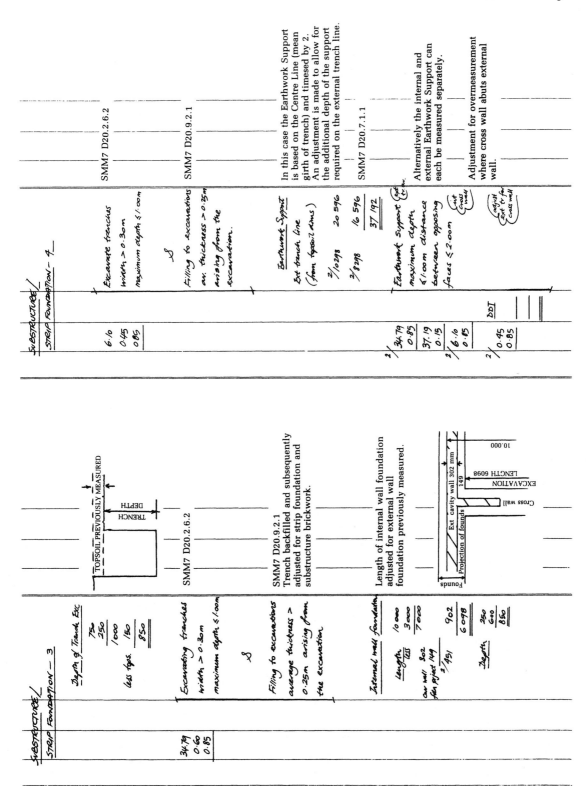

SMM7 D20.2.6.2

SMM7 D20.9.2.1

In this case the Earthwork Support is based on the Centre Line (mean girth of trench) and timesed by 2. An adjustment is made to allow for the additional depth of the support required on the external trench line.

SMM7 D20.7.1.1

Alternatively the internal and external Earthwork Support can each be measured separately.

Adjustment for overmeasurement where cross wall abuts external wall.

SMM7 D20.2.6.2

SMM7 D20.9.2.1
Trench backfilled and subsequently adjusted for strip foundation and substructure brickwork.

Length of internal wall foundation adjusted for external wall foundation previously measured.

PLAN OF EXTERNAL WALL AND
CENTRE LINE ADJUSTMENT AT
CORNERS

½ wall thickness

External wall length

External wall width

C.L. of ext wall

Perimeter length based on external wall dimensions.

Adjusted to deduct half wall thickness from each length and width at corners.

E.g. for external skin of wall only

∴ If wall = 100 mm th adjust each corner by $\left(\frac{100}{2}\right)$ 50 mm } ×2

once from length
once from width

Normally four corners

∴ deduct $^{4/2}/50 = 400$ mm from perimeter length.

Cross wall length based on previous waste calculation (see page 3 of this take-off).

SUBSTRUCTURE
STRIP FOUNDATION – 6

Blockwork Q.s

Cavity Ext. block wall C.L.

External wall dims.   36 000

$\left(\frac{100}{2}\right)$ 50 4/2/50   400

less 4/2/50

C.L. ext skin = 35 600

Cavity int. block wall C.L.

Ext. wall dims.   36 000

100
50
75 $\left(\frac{150}{2}\right)$
225 4/2/225 / 800

less 4/2/225

C.L. int skin = 34 200

Height of Blockwork

Ext. skin 750
less 3/75 150
(2 courses brk) 600

int skin 750
add 38
(½ course brk) 788

Int wall blockwork length

excavated length 6098
add 2/149 298
                6396

SUBSTRUCTURE
STRIP FOUNDATION – 5

		Surface treatments compacting bottoms of excavation (ext wall)	SMM7 D20.13.2.3
34.79	0.60		NB If blinding with concrete to bottoms of excavation measurement required in accordance with SMM7 E10.
	0.45		
6.10	0.45		
34.79	0.60	Plain in-situ concrete 20 kN/m² – 20mm aggr. poured on or against earth or unblin (int wall) old hardcore	SMM7 E10.1.0.0.5
	0.25		
6.10	0.45	∫	
	0.25		
		Deduct	
		Filling to excavation as before described	SMM7 D20.9.2.1
		∫	
		Add	
		Disposal excavated material off site,	SMM7 D20.8.3.1

SUBSTRUCTURE
STRIP FOUNDATION – 0

Cavity fill & cavity form

C.L. as previous

Cavity Fill Height 750

less 2/75   150
          600

C.L. calcs for cavity

External wall   36 000

100 (50/2)
35 (50/2)
125   less 4/125   1 000
          35 000

Cavity form parapet   750
add   150
       900

35.00 0.90	Forming cavities in hollow walls 50mm wide including triangular wall ties/m²

SMM7 F30.1.1.1

35.00 0.05 0.60	In situ concrete (weak mix) filling to hollow wall thickness < 150mm

SMM7 E10.8.1.0

SUBSTRUCTURE
STRIP FOUNDATION – 1

Ext wall height   600
          75
        675

35.60 0.60 6.40 0.68	Walls 100mm thick dense concrete block (ext wall) laid stretcher bond in c.m. (1:3) (int wall)

SMM7 F10.1.1.1

No need to describe work as 'vertical' (despite SMM7 F10.1.*.1) since Definition Rule D3 states that 'work is deemed vertical unless otherwise described'.

Facework to dpc at

C.L. as before

Height above g.l.   130
2 courses below gl 2/75   150
          300

35.60 0.30	Walls facework one side 103mm thick laid stretcher bond in c.m. (1:3) with a rubbed joint each

SMM7 F10.1.1.1

Alternatively might have been measured up to dpc level as blockwork and adjusted here for facing bricks in lieu of blocks.

34.20 0.74	Walls 150mm thick dense concrete block laid stretcher bond in c.m. (1:3)

SMM7 F30.1.1.1

SUBSTRUCTURE

STRIP FOUNDATION – 10

Work to Structural Floor

TO TAKE:– Oversite excavation *
    Surface treatments
    Beam & Strip Floor

* measured with tops exc.

Ignore dims
length 10298
less points 2/100/200
    8098

width 8000
less points 3/600/200/200
    6800

8·10		Surface treatments
6·80		compacting ground
	DDT	
6·10		8 (cross wall)
0·45		
4·00		Surface treatments (hand)
3·00		applying herbicides
		PC200 at the rate
		of 25g/10m²

SMM7 D20.13.2.1

SMM7 D20.13.1.0

Alternatively, surface treatments may be given in the description of any superficial item.

(Measurement Rule D20.M17)

SUBSTRUCTURE

STRIP FOUNDATION – 9

Adjust b/fill for our wall and not cross wall
Height of walling in exc.
    750
  less tops 150
    600

34·79	C.L. Our wall. ad ph. length ...
0·30	**Deduct**
0·60	Filling to excavation a.b.d.
6·40	
0·10	**Add**
0·60	Disposal excavated material off site.
35·60	Damp Proof Course.
0·10	
2/34·20	Damp proof course ≤ 225mm horizontal
0·15	single layer lapped
2/6·40	150mm at joints
0·15	bedded in c.m. (1:3)

Trenches previously entirely backfilled now adjusted for cavity wall.

Dimensions previously established as waste calculations and therefore no need to repeat this.

SMM7 D20.9.2.1

SMM7 D20.8.3.1

NB Crosswall blockwork previously measured through internal trench backfill would normally require adjustment.

SMM7 F30.2.1.3

SECTION THROUGH EXTERNAL WALL FOUNDATION SHOWING DPC

BEAM

Dpc above and below beam on internal foundation wall

SUBSTRUCTURE
STRIP FOUNDATION – 12

Structural Floor

Rackham House Floor
all as detailed on drawgs.
RHF/SF8507; 150mm deep
reinforced pre cast concrete
'I' beams with 100mm
dense concrete blockwork
infill

Floor units (Bay 1)
4149 mm long
| 6.40 |
| 4.15 |

Ditto (Bay 2)
3694 mm long
| 9.40 |
| 3.69 |

Slurry screed brush
applied horizontal
surfaces of beam &
block infill floor.
| 6.40 |
| 4.15 |
| 9.40 |
| 3.69 |

SUBSTRUCTURE
STRIP FOUNDATION – 11

Length of floor beams
Bay 1
4000
less
½ bk wall 103
cavity 50   153
3847
add
Cavity wall
= int wall
bearing   302
Bay 1 Length 4149

Bay 2
4200
less
½ bk wall 103
cavity 2/153   306
Bay 2 Length 3894*
*50mm o'sail ext cavity wall.

Bay 1   less 10000
3000
7000
604
63%
less 2/302   63%

Bay 2   1000
604
less 2/302   93%

External wall and foundation plan

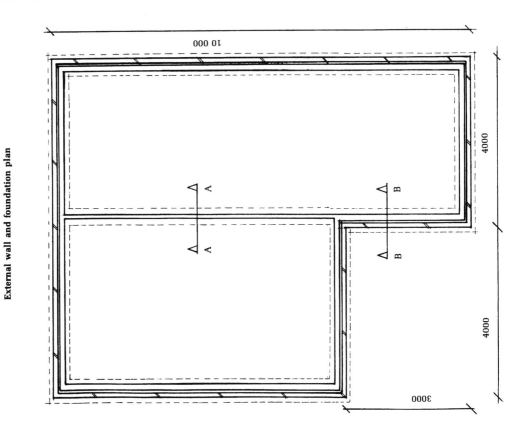

10 000

4000

4000

3000

SUBSTRUCTURE

STRIP FOUNDATION – 13

Air bricks & floor
ventilators.

Girth of ext wall
2/10 000   20 000
2/8 000   16 000
           36 000

Vents @ 1.80 m centres
∴ 36 000 / 1.800 = 20 nos +1
9.84 vents = 21

Air brick; terra-cotta
built into 225 × 75 mm
opening in 103 mm th
facing brickwork wall
a.w.p.

8

Plastic two-piece
adjustable telescopic
floor ventilator unit
built into cavity wall
comprising 103 facework
50 cavity and 150
inner skin a.w.p.

21

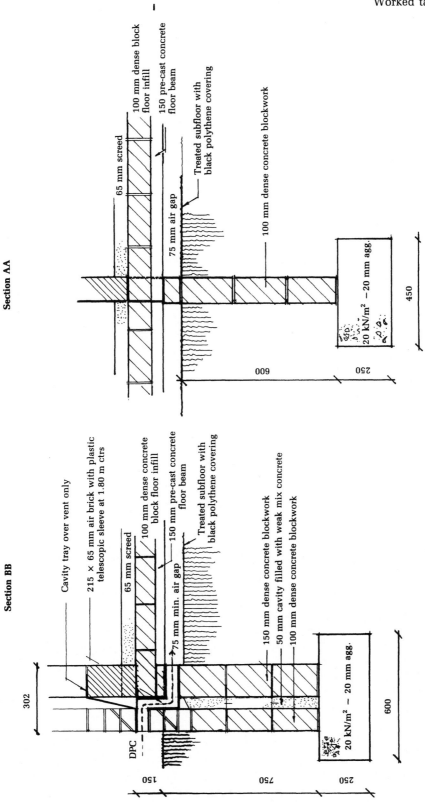

Section AA

Section BB

6 Masonry work

6.1 Introduction

In previous editions of the Standard Method the term 'Masonry Work' had been used exclusively to describe the trade of the stone mason together with the dressing, coursing and laying of stone. The present edition of the Standard Method adopts this same term to embrace the trades of brickwork and blockwork together with cast and natural stonework.

These are classified as follows:

F Masonry Work	F10	Brick/block walling
	F11	Glass block walling
	F20	Natural stone rubble walling
	F21	Natural stone/ashlar walling/ dressing
	F22	Cast stone walling/dressing
	F30	Accessories/sundry items for brick/block/stone walling

When measuring brick and block walling it is always advisable to follow a set order (this would normally replicate the sequence of construction) and to provide a take-off list. Under normal circumstances, masonry work below dpc level would be included as part of the substructure measurement, even though there is no requirement in SMM7 to distinguish between work in substructures and work in superstructures. Nevertheless, the general approach to measurement will require the separation of work between substructures and superstructures. On large projects where there are a number of different types of walling, it is helpful to colour code the drawings accordingly. The more usual alternatives for walling will be determined by location (substructure/superstructure, internal/external) or function (structural/nonstructural).

With the exception of SMM7 F30, the principal unit of measurement for Masonry Work is the square metre. The two dimensions needed to provide this area are the centre line length and the height of the walling. The description will identify the walling thickness, the type of brick or block, the bond, the type of mortar and the pointing (SMM7 F10.S1). Rather than repeat these details in every description, a heading can be established to include the majority of this information. The subsequent descriptions need only make reference to the wall's thickness, its plane (when not vertical) and whether it is facework on one or both sides. The thickness of walling given in the description can be recorded in one of two ways: either by reference to the standard brick sizes or by stating the sectional width of the walling. To avoid any doubt, some surveyors prefer to give both (refer to Brick/block walling descriptions, section 6.2).

The majority of cavity walls are constructed with an outer leaf in brickwork, a cavity, which may or may not be insulated, and an inner leaf of blockwork (usually with some insulation properties). Each of these layers will require measuring as a separate item and each will require a separate waste calculation in order to establish its centre line (SMM7 F10.M1). The purpose and location of walls will largely determine the type of brick or block used and the finish on the face of the exposed wall. The following terms are commonly employed in the construction industry and by the Standard Method to distinguish between the various finishes of brickwork.

- Common brickwork: walling which does not need to include any facing bricks or pointed finish. Walls of this type will normally remain unseen once building work is completed. They are often used below ground level, or in situations where they will subsequently be covered.
- Facework: walling which will be built such that the exposed or 'facing' part is left with a pleasing finish. Care would be taken to avoid using chipped or irregular shaped bricks and the mortar joints would be pointed on completion.

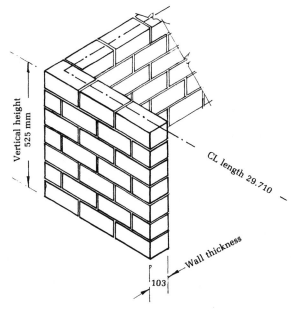

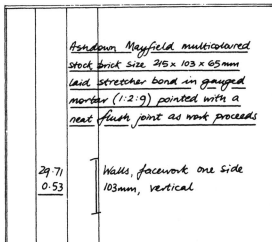

Figure 6.1 Masonry work measured in square metres.

6.2 Brick/block walling descriptions

Since the cost of facing bricks is significantly more than that of common bricks, it is necessary in the descriptive part of the measurement to distinguish between these two basic types of walling. SMM7 F10.*.1, 2 and 3 provide the alternative descriptions (common brickwork (no finish), facework one side, and facework both sides).

There are four main classifications for brick/block

walling (SMM7 F10.1, 2, 3 and 4) and a host of other associated classes, ranging from projections to surface treatments (SMM7 F10.5 to F10.26). All walling must be allocated for descriptive purposes to one of these classes. Mention should also be made of the shape or plane that the finished walling will take: vertical, battering (sloping walls with parallel sides) or tapering on one or both sides (walls of diminishing thickness). There is an example of conflicting information with regard to this last requirement since Definition Rule D3 states that 'work is deemed vertical unless otherwise described'.

Finally, in some special circumstances, the description can be suffixed with the appropriate classification from SMM7 F10.1.*.*.1 to 4. Either as part of the normal description or perhaps as a general heading to the masonry brick/block walling trade, details must be provided of the items identified in SMM7 F10.S1 to S5. As noted previously, it will prevent repetition if the latter approach is adopted. As is the case with all trades in SMM7, it will be necessary to identify the scope of the work in accordance with the requirements of clause SMM7 F10.P1.

An alternative approach could be adopted where a heading is established on the sheet of dimension paper (figure 6.3). This would save time where there were a number of descriptions relating to the same facing brickwork.

A similar procedure (figure 6.4) can be adopted when measuring block walling.

In these examples, all the dimensions are assumed and in the interest of brevity waste calculations have been omitted.

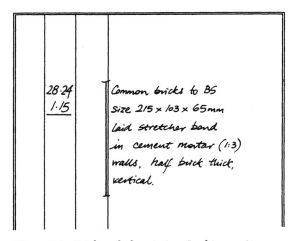

Figure 6.2 Brickwork description. In this case it would appear that the brickwork will remain unseen (probably in foundations) since no detail is given of any facework. This particular wall will be 'left as laid'.

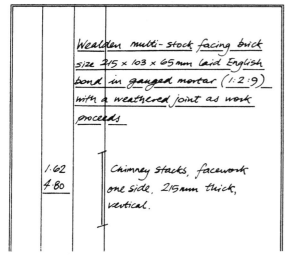

Figure 6.3 Using a heading to establish a common description. It can safely be assumed that this brickwork will be on view since the description makes reference to 'facework one side' and the brick being used is a facing brick.

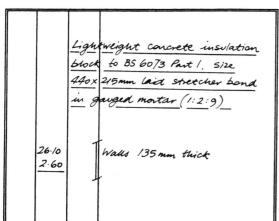

Figure 6.4 Use of a heading is made here with regard to blockwork. The description implies the work will not be seen on completion. A description like this would be used for the inner skin of a cavity wall. Note that the external skin and the cavity would each require separate measurement.

6.3 Approach to measurement

6.3.1 Cavity walls

Each skin of a cavity wall is measured separately in square metres on its individual centre line. Forming cavities in hollow walls is measured in the same way, stating the width of the cavity in the description together with the type, size and spacing of wall ties. If the cavity includes any rigid

sheet insulation, this can be incorporated as part of the description of forming cavities, stating type, thickness and method of fixing.

Openings in walls for windows and external doors are, at this stage, ignored. An adjustment will be made at the time of measuring the window or door, against this initial over-measurement.

When measuring the external walling to traditional gable-ended properties, the stages for the measurement of masonry work would be as follows:

Stage 1 — Measure a box (or rectangle) based on the centre line × height to soffit level, for each skin of the wall and the cavity.

Stage 2 — At eaves measure brickwork to close the cavity along the length of the soffits — this can be booked in square metres in accordance with

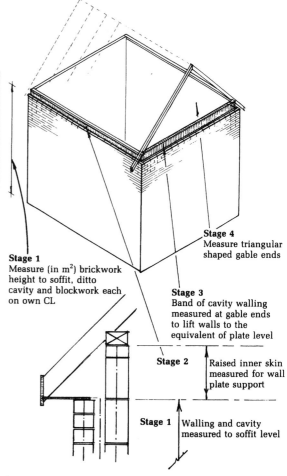

Figure 6.5 Stages of measurement for gable-ended buildings.

F10.1 or it may be described as closing cavities horizontally and measured linear (F10.12). Alternatively, the cavity wall may be left open at eaves level and the blockwork inner skin raised independently to support the wall plate (details will vary depending on circumstances). The wall plate will be measured as part of the structural timber Work Section.

Stage 3 — Before measuring the triangular section of gable
and end cavity walling at either end of the dwelling,
Stage 4 it will be necessary to include a narrow band of cavity work to lift these end sections to the same height as the brickwork which was measured to close the cavities at soffit level (Stage 2).

6.3.2 Partition walls

To complete the measurement of masonry work associated with domestic dwellings it will be necessary to book dimensions for internal partition walls. Care should be taken to distinguish between stud walling and other internal walling systems which must be measured under a different set of rules. Under normal circumstances the floor to ceiling height will be consistent on each storey and one of the techniques shown in figure 6.6 may be used.

Both brickwork and blockwork share the same classification table and set of rules. As with cavity walling, openings for internal doors will be ignored, an appropriate adjustment being made later.

6.3.3 Other classifications

The previous pages have identified, at an introductory level, how to set down dimensions and descriptions for masonry work. However, there are a number of other categories of brickwork/blockwork which require explanation. For ease of reference these are given under their SMM7 F10 coded classification.

F10.5 Projections Includes attached piers, oversailing courses and plinth courses. For attached piers the height of the projection is measured in linear metres, stating its width and depth together with its plane in the description (figure 6.7). Where the length of a projection exceeds four times its thickness then it is defined by SMM7 as being a wall and measured accordingly (SMM7 F10.D9) (figure 6.8). Isolated piers/ walls, oversailing courses and plinth courses are measured in the same fashion using the same clause.

F10.6 Arches Measured on the mean girth on face in linear metres stating the height. The number of identical

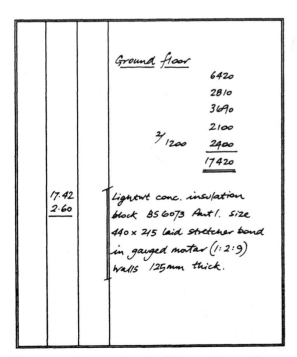
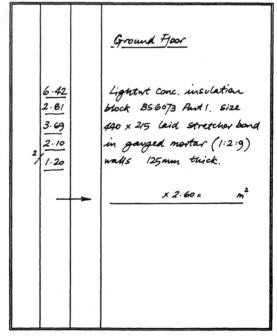

Figure 6.6 An example take-off for partition walls.

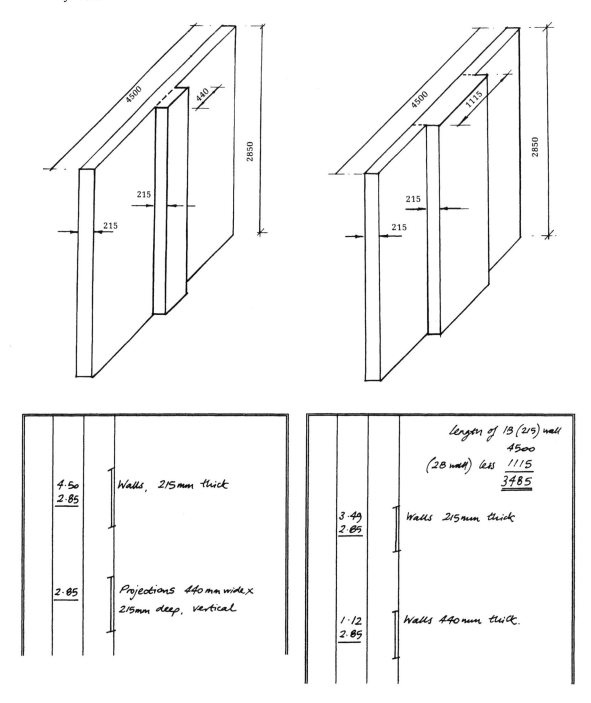

Figure 6.7 An attached pier designated as a projection: the length of the pier is less than four times its width and is therefore measured as a projection.

Figure 6.8 An attached pier designated as a wall: the length of the pier exceeds four times its width and is therefore measured as a wall.

arches, together with the width of the exposed soffit and the arch shape (e.g. segmental/flat), must also be included in the description (figures 6.9 and 6.10).

F10.12 Closing cavities Most likely to occur where windows and external doors are measured and in the normal course of events these would be included as part of the window or door opening adjustment. These are booked in linear metres, stating the width of the cavity and the method of closing together with the plane (e.g. vertical, horizontal) (figure 6.11). It is often necessary and convenient to include the measurement of any associated dpc with the measurement of closing cavities. SMM7 F30.2 should be consulted regarding the measurement of vertical damp proof courses.

F10.13 Facework ornamental bands All are measured in linear metres stating whether flush, sunk or projecting, together with the width of setback or setforward and the plane. An option is available to measure projections as 'extra-over' the work in which they occur (figure 6.12).

F10.14 Facework quoins A similar approach is adopted for the measurement of facework quoins, the mean girth

of the quoin being stated in the description and the vertical angle length providing the booked dimension. The provisions of this clause would only be applied where the quoin brickwork differs from the facings used in the body of the walling (figure 6.13).

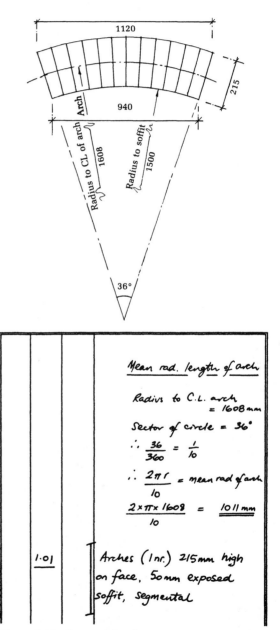

Figure 6.10   Segmental arch.

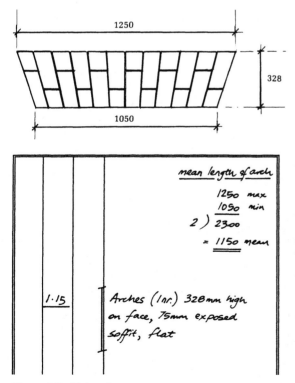

Figure 6.9   Flat arch.

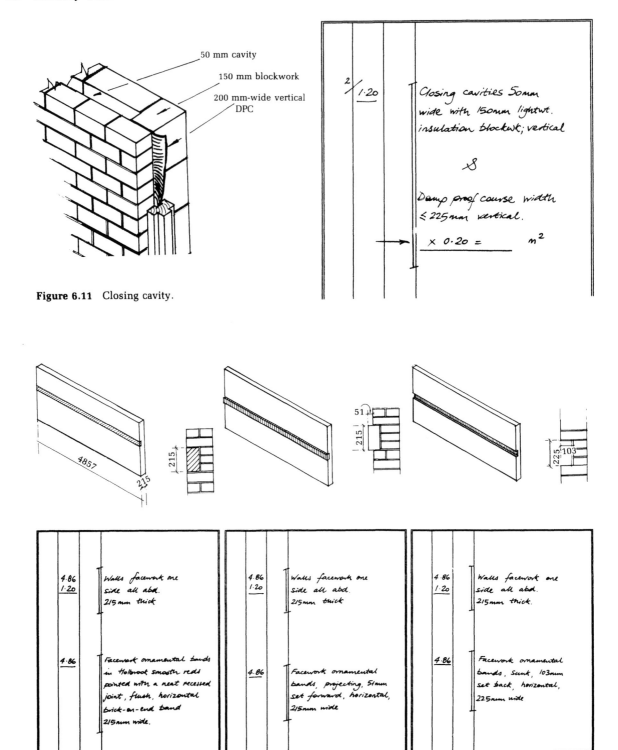

Figure 6.11 Closing cavity.

Figure 6.12 Ornamental bond facework.

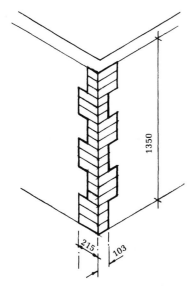

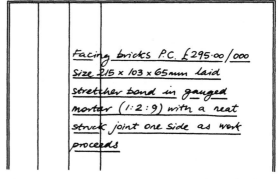

Figure 6.14 A prime cost sum to cover facing brick supply.

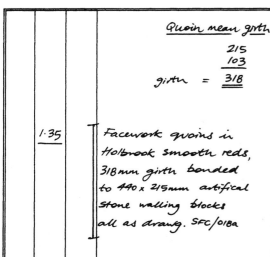

Figure 6.13 Facework quoins.

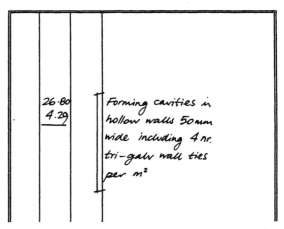

Figure 6.15 Forming cavities. Additionally, where rigid sheet cavity insulation is used, this can be included in the description.

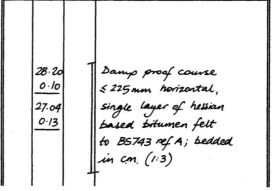

Figure 6.16 Damp proof course. No allowance is made in the booked dimensions for laps. Dimensions are based on the centre line length multiplied by the width of walling.

F10.15 to F10.24 There are also rules for the measurement of facework sills, thresholds, copings and steps (F10.15 to F10.18), all of which are reasonably self-explanatory. Together with individual facework items such as keyblocks, corbels and cappings (F10.19 to F10.24) these items can be measured in linear metres or as enumerated items, whichever is the more appropriate.

On occasions, the architect (or client) may not have selected a facing brick at tender stage. In such circumstances, a Prime Cost Sum may be included to cover the cost of the supply of the facing bricks. Since these are purchased in units of 1000, the price is usually expressed in the same fashion, as in figure 6.14.

6.3.4 Sundry items

This section of SMM7 covers the more common sundry items associated with the work of the bricklayer. In the majority of domestic buildings this would include forming cavities, damp proof courses, flue linings, air bricks and lintels (proprietary items). Most of these are self-explanatory, and sample entries are shown in figures 6.15 (forming cavities, F30.1), 6.16 (damp proof courses, F30.2) and 6.17 (metal lintels, F30.16). For precast concrete lintels, see SMM7 F31.

6.4 Worked take-off example

A worked example of a take-off for structural walls is shown on the following pages (take-off sheets: pp. 69−72; drawings: p. 73).

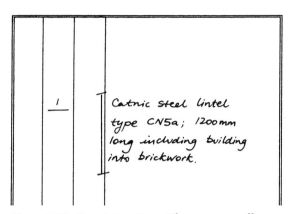

Figure 6.17 Proprietary items. There are normally a number of lintels and these are more economically measured by setting up a heading. The height of the lintel will normally coincide with a standard brick/block course height. Where this displaces a full brick or block course, an adjustment should be made; where it does not, no adjustment is necessary (SMM7 F10.M3).

STRUCTURAL WALLS – 2

	3.09	Walls facework one side half brick thick (103) in Funtley red facing brick laid stretcher bond in gauged mortar (1:2:9) pointed with a neat rubbed joint as work proceeds. *(gable ends / gable raising)*
	4.88	
2/	8.30	
	0.16	
2½/	8.30	
	3.50	

SMM7 F10.1.2.1.*

NB No need to state work as 'vertical' since Definition Rule F10.D3 deems work as vertical* unless otherwise described.

Centre line of cavity

ext girth 30 500
less 103
(50/2) 25
128   4/ /128 1.024
      29 476

Centre line calculations based on external wall perimeter length previously established.

	29.48	Forming cavities in hollow walls 50mm wide inc. 4 nr tie-galv wall ties per m² *(gable raising)*
	4.88	
2/	8.30	
	0.16	
2½/	8.30	
	3.50	

Cavities measured as an area based on centre line of cavity × vertical height.

SMM7 F30.1.1.1
Possible to include cavity insulation in the description; state type thickness and method of fixing.

STRUCTURAL WALLS – 1

TAKE-OFF LIST:

Facing brick } House ext.
Cavity   } walls
Insulation b/k

Facing brick } Garage
Solid blk   } walls

Adjustment of facework for garage

Projections

Internal g.f partitions

Measurement taken over all openings; adjustment for windows and doors made later (ref. chapter 10 Standard Joinery).

Initial measurement for two-storey dwelling; attached garage block measured separately later.

PLAN

ATT. GAR.

DWELLING

External girth brickwork
2/ 6950   13 900
2/ 8300   16 600
ext girth = 30 500

Centre line of ext skin
less 4/ ½/ 103   412
C.L. ext skin   30 088

Brickwt tot to soffit
dpc to f.floor   2400
floor joist   225
joist to soffit 2250
                 4875

Centre line calculations based on external wall perimeter length.

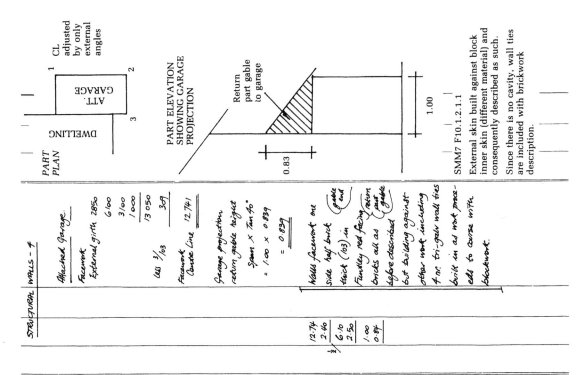

PART PLAN

CL adjusted by only external angles

PART ELEVATION SHOWING GARAGE PROJECTION

Return part gable to garage

SMM7 F10.1.2.1.1

External skin built against block inner skin (different material) and consequently described as such.

Since there is no cavity, wall ties are included with brickwork description.

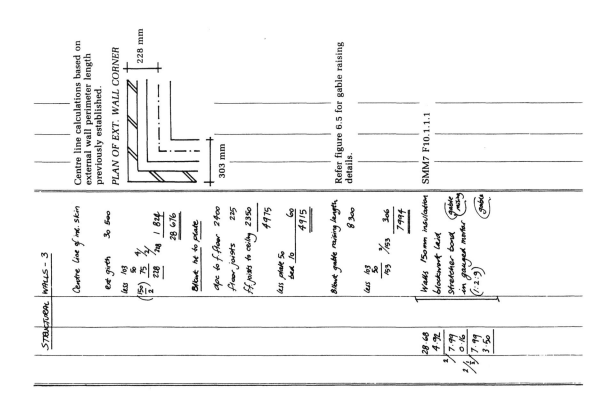

Centre line calculations based on external wall perimeter length previously established.

PLAN OF EXT. WALL CORNER

Refer figure 6.5 for gable raising details.

SMM7 F10.1.1.1

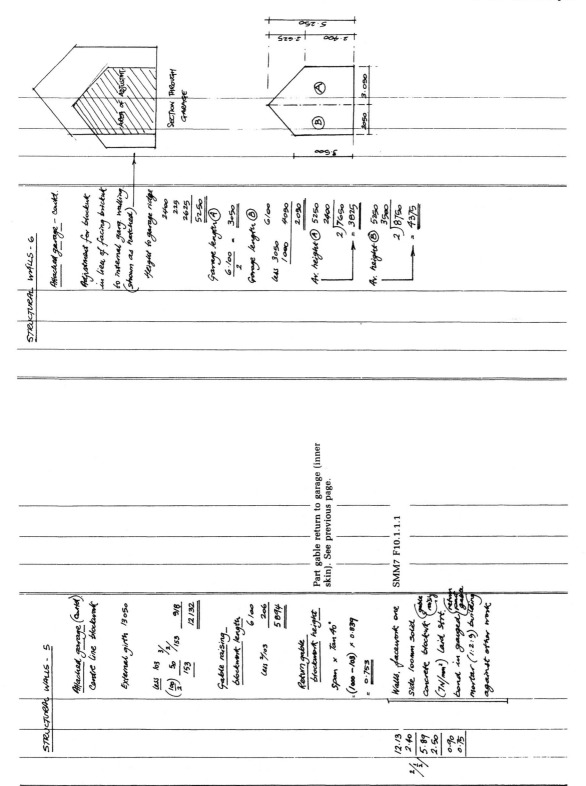

STRUCTURAL WALLS - 6

Attached garage - contd.

Adjustment for blockwork in lieu of facing brickwork to internal garg walling (shown as hatched)

Height to garage ridge
2400
225
2625
5250

Garage length (A)
6100/2 = 3050

Garage length (B)
6100

less 3050   4050
     1050   2050

Av. height (A) 5250
                2400
              2)7650
              = 3825

Av. height (B) 5250
                3500
              2)8750
              = 4375

STRUCTURAL WALLS - 5

Attached garage (contd)
Cavity line blockwork

External girth 13050

less 3/2/153   9/8
(100/2) 50/153   12/32
        153

Gable raising -
blockwork length   6100
less 7/03   206
                   5894

Return gable blockwork height
Span × Tan 46°
= (1000 - 103) × 0.839
= 0.753

2/1/2/	12.13		
	2.40		
	5.89		
	2.50		
	0.90		
	0.75		

Walls, facework one side 100mm solid concrete blockwork (7N/mm²) laid strt bond in gauged mortar (1:2:9) building against other work

Part gable return to garage (inner skin). See previous page.

SMM7 F10.1.1.1

STRUCTURAL WALLS – 8

Internal struct walls

Lengths 3650
100
3944
2000
100
4244
14938

Height struct walls 2400
through joist 225
2625

14.94
2.63

Walls 100mm thick Lightweight concrete blockwork laid stretcher bond in gauged mortar (1:2:9)

SMM7
F10.1.1.1

14.94
0.10

Pitch polymer single layer dpc width ≤225mm horiz. bedded in gauged mortar (1:2:9)

SMM7 F30.2.1.3

Plan lengths transferred from drawing, totalled in waste and transferred to dimension column. Assume first floor partitions are studwork.

Coverage Rule F10.C1 does not specifically mention joist filling as deemed included labour item – this is assumed.

STRUCTURAL WALLS – 7

Attached garage – contd.

Ⓐ

Deduct
3.05
3.83
2.05
4.38

Walls facework one side half brick thick Ⓑ in Fletton reds (103) in Fletton reds a.b.d.

SMM7 F10.1.2.1.1

Add

Walls facework one side 100mm Solid Conc 7N/mm² laid Stretcher bond in gauged mortar (1:2:9)

SMM7 F10.1.2.1.1

Attached piers

2/ 2.40

Projections 215mm wide x 103mm deep; vertical bonding to other work

SMM7 F10.5.1.1

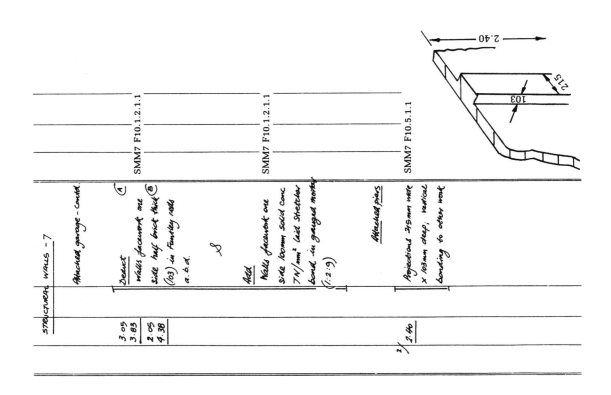

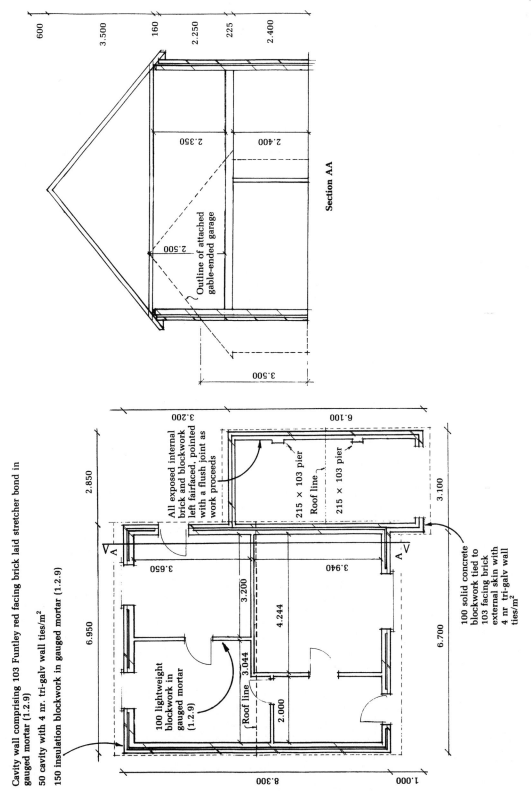

Cavity wall comprising 103 Funtley red facing brick laid stretcher bond in gauged mortar (1.2.9)

50 cavity with 4 nr. tri-galv wall ties/m²

150 insulation blockwork in gauged mortar (1.2.9)

Section AA

600
3.500
160
2.250
225
2.400

2.350
2.400
2.500

3.500

Outline of attached gable-ended garage

All exposed internal brick and blockwork left fairfaced, pointed with a flush joint as work proceeds

215 × 103 pier
Roof line
215 × 103 pier

100 solid concrete blockwork tied to 103 facing brick external skin with 4 nr tri-galv wall ties/m²

100 lightweight blockwork in gauged mortar (1.2.9)

Roof line

3.200
6.100
2.850
3.100

3.650
3.940

3.200
4.244
3.044
2.000

6.950
6.700

8.300
1.000

External Walls. Two-storey detached dwelling with attached garage

7 Structural timber

7.1 Introduction

Timber has been used as a structural component in buildings since time immemorial. Compared with other structural materials (concrete and steel) it is not particularly strong, but this is compensated by its light weight which allows for large timbers to be employed without excessive increase in the weight of the structure.

Increasingly structural timber components are prefabricated and for pitched roofs of a domestic scale there are few options: trussed rafters dominate. A comparison of cost between a traditional cut roof and a trussed rafter roof shows little difference in the cost of supply; however, the trussed roof is reputed, on average, to be four times as fast to erect (*BRE Digest* 147).

The process of prefabrication was taken one stage further with the introduction of timber-framed housing which presently accounts for 20% of the new build housing market. Despite its rather poor public image, if properly erected it can offer significant savings in time and a higher level of thermal insulation than could normally be achieved with traditional building. It is particularly popular in the Scandinavian countries where it satisfies the most stringent thermal regulations in Europe.

7.2 Units of measurement and standard sectional timber sizes

The unit of measurement for structural timber naturally falls between enumerated items (trusses, etc.) and linear metres (joists, rafters, studs, etc.). In the case of the latter the description must include the cross-sectional dimensions of the timber being measured, whilst the former must fully describe the trussed rafter and is undoubtedly best communicated with the aid of a dimensioned bill diagram.

Timber used in the construction industry is sawn at the mill into standard sectional sizes which are known as nominal or basic sizes (see table 7.6). The process of sawing leaves an irregular or rough surface on the face of the timber which is generally referred to as 'sawn'. Most structural timbers are supplied sawn and the Standard Method deems that all timber is expressed in its sawn or nominal size (SMM7 G20.D1). Where a smooth or planed wrot finish is required, the dimensioned description should be suffixed with the word 'finished' or 'fin'. It is unlikely that this will apply to any structural timbers (other than trusses) although some first fix items, e.g. fascia and soffit boards, are grouped for measurement purposes within this structural section of the Standard Method. On the few occasions where a wrot finish is required SMM7 clause G20.19 allows for a linear measurement stating in the description the width, or girth, on which a planed finish is needed.

A more detailed explanation of the sizing and measurement of non-structural timber components can be found in chapter 10.

7.3 Approach to measurement

Drawn details identifying the scope and location of the work should be provided to satisfy the conditions of SMM7 G20.P1. It is also important to include details of the kind and quality of timber together with the method of fixing (if other than nailed) and any surface treatments applied as part of the production process. These, along with other similar details, are listed under the heading Supplementary Information (SMM7 G20.S1 to S9). To save unnecessary repetition these may be included in a general heading to the take-off, as shown in figure 7.1.

The classification table given in SMM7 G20 discriminates between prefabricated structural timber items delivered to site ready for inclusion in the works (G20.1 to 5) and traditional cut timbers such as rafters and joists (G20.6 to 8). Many projects will include a combination of both and it is obviously important to determine which is which.

Figure 7.1   Take-off heading giving timber specification.

A number of techniques are in common use to establish the number and length of structural timber components and a brief résumé of these follows.

7.3.1 How many joists?

Drawn information provided by architects rarely includes a floor or roof plan which identifies the actual number of joists, rafters or trusses required. Instead, the measurer must establish these from the other information available on the drawing. This will include the spacing centres of joists or rafters (commonly 400, 450 or 600 mm) together with external or internal plan dimensions and an indication on the drawing of the span direction (normally the shorter of the room plan dimensions), which is usually represented by a directional arrow ($\leftrightarrows$).

There are five stages in establishing the number of joists:

(1)   Find the internal length of the room.
(2)   Find the centre to centre dimension of the first and last joists.
(3)   Divide the dimension in (2) by the distance between joist centres to establish the number of *spaces*.

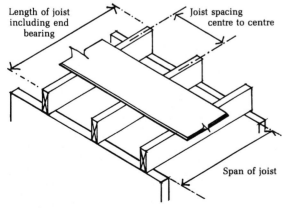

Figure 7.2   Joist measurements.

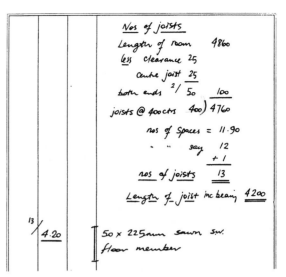

Figure 7.3   Establishing the number of joists.

(4)   Round the answer up to the next whole number (on rare occasions the answer will already be a whole number).
(5)   Having found the number of *spaces*, add one to achieve the number of *joists*.

7.3.2 How many rafters?

A very similar approach should be adopted to establish the number of rafters in a hipped roof.

The inclusion of hips and valleys is ignored when measuring common rafters and no distinction is made between hipped and gable-end roofs, since the total length of jack-rafters at a hip or valley is nominally equal to the length of the corresponding common rafters in a gable roof.

The combined length on plan of each pair of jack-rafters is equivalent to one common rafter. But it will be necessary to take one extra rafter at each end on the same line as the ridge as detailed in figure 7.4.

The waste calculation used to identify the number of joists can also be used for establishing the number of trussed rafters, cut rafters in a gable roof or studs in partition walls.

7.3.3 How long?

A further waste calculation will be necessary to establish the length of structural timbers. In the case of floor or ceiling joists this is reasonably straightforward and simply requires the addition of end bearing to the clear span dimension (see figures 7.2 and 7.3).

Alternatively, where the joist is suspended in joist hangers the measured length of timber would simply be the clear

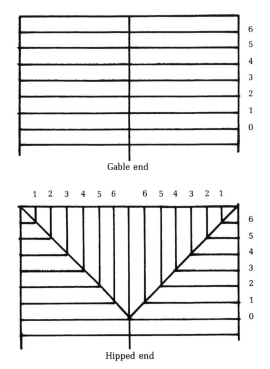

Gable end

Hipped end

Figure 7.4 Numbering rafters in gable- and hipped-end roofs.

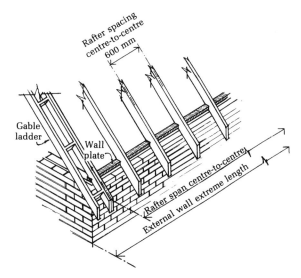

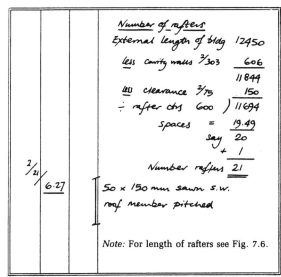

Figure 7.5 Establishing the number of rafters.

the clear span. In this case an inclusion must also be made for joist hangers and these will be enumerated in accordance with SMM7 G20.21. The labour of notching the joist into the joist hanger is deemed included (SMM7 G20.C1 and C3).

A similar approach may be adopted for the vertical studs in a stud partition wall, with appropriate adjustment for head and sill timbers.

Some allowance may also be necessary for jointing the vertical studs to the head and sill, although this may also be achieved with a metal fixing (SMM7 G20.20 to 28).

The length of rafters can be established in a number of ways, shown in figure 7.6.

Table 7.1 gives the secants of common roof slopes.

Length of hip and valley rafters Figure 7.7 shows a plan view of a hipped roof. The length of a hip or valley cannot be scaled direct from plan since the length required is the length on slope. The following method is used:

(1) Set out the roof as above, at right angles to the hip.
(2) Scale off the vertical height of the roof along this line.
(3) Join the two lines to form a right-angled triangle and scale length of hip from drawing.

Complicated roof shapes These should be subdivided into simple units, measuring the principal unit overall and adding the projections (figure 7.8).

7.4 Presentation in the Bill of Quantities

The descriptive part of the measurement was partly covered in the introduction to this chapter. For traditional, rather than prefabricated components, the Standard Method uses the classifications given in table 7.2. These are based on the function and location of the structural timber member.

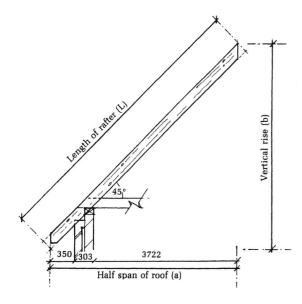

Table 7.1 Secants of common
roof slopes

Roof slope	Secant
17½°	1.049
22½°	1.082
30°	1.155
35°	1.221
37½°	1.260
40°	1.305
45°	1.414

Length of rafter can be found by using one of the following
methods:

1. Scale length from section
2. Use Pythagoras theorem $L = \sqrt{(a^2 + b^2)}$
3. Use Natural Secant $L = a \times \sec 45°$

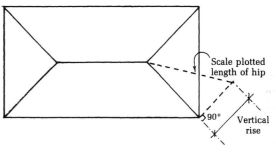

Figure 7.7   Determining the length of a hip or valley.

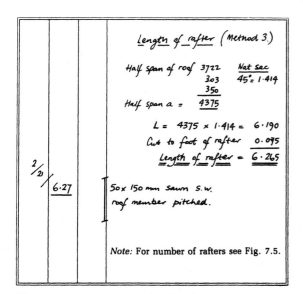

Figure 7.6   Determining rafter length.

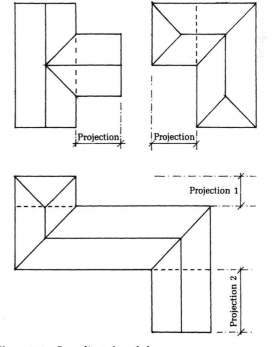

Figure 7.8   Complicated roof shapes.

The length of the timber being measured is recorded in
the dimension column, and the timesing column is used to
multiply this by the number of times the particular
component occurs. The Bill of Quantities will, of course,

Table 7.2 SMM7 classifications of traditional structural timber components

SMM7 ref.	SMM7 description	Components
G20.6	Floor members	Floor joists Timber beams
G20.7	Wall or partition members	Studwork Heads and sills Noggins and struts
G20.8	Plates	Wall and floor plates Bearers
G20.9	Flat roof members	Roof joists Timber roof beams
G20.9	Pitched roof members	Rafters, hips and valleys Purlins, binders, bracing and struts Ridge board and ceiling joists

record only a single length accompanied by a description which identifies the type of timber, its location or function and its cross-sectional dimension.

A completed description for, say, ceiling joists might be presented as in figure 7.9. A similar layout and presentation should be adopted for other structural timber items. Note the use of signposting to identify ceiling joists. The Bill of Quantities will group all structural timber of this sectional size in a pitched roof together and there will be no obvious way to distinguish a rafter from a ceiling joist.

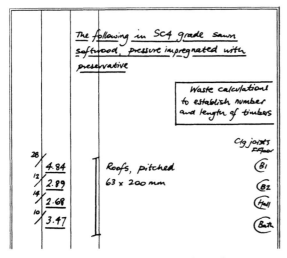

Figure 7.9 A completed description for ceiling joists.

The take-off lists given in tables 7.3, 7.4 and 7.5 are offered for general guidance only. Since every situation will vary, it is unlikely that the construction will exactly match this.

Table 7.3 Timber intermediate floor take-off list

Item	Unit	SMM7 ref.
Wall plates	m	G20.8
Floor joists	m	G20.6
Timber beams	m	G20.6
Joist hangers, metal straps and connectors	nr	G20.10 and 20
Joist strutting (herringbone or solid)	m	G20.10
Insulation	m²	P10.2
Boarding	m²	K20.2
Opening adjustments	—	various

Table 7.4 Stud partition take-off list

SMM7 ref.	Item	Unit
G20.7	Partition members: includes studs head/sill noggins	m
G20.22	Metal connectors (where no allowance for jointing made in above)	nr

Table 7.5 Traditional cut pitched roof take-off list

SMM7 ref.	Item	Unit
G20.8	Plates	m
G20.9	Rafters Ceiling joists Hips and valleys Purlins, hangers and struts	m
G20.14	Gutter boarding	m/m²
G20.17	Sprockets or	nr
G20.13	Tilting fillet	m
G20.20	Ties to walls	nr
G20.8 G20.12	Tank support platforms	various
K20.2	Loft boarding	m²
P10.2	Insulation	m²

7.4.1 Trussed rafters

The rules for this section have been drafted on the assumption that the majority of structural timber components will now be shop processed using machinery rather than cut and fixed on site. SMM7 General Rule 9 makes specific reference to the treatment of composite items.

The measurement of trussed rafters neatly fits this category and is best described by reference to a bill diagram or dimensioned diagram (SMM7 General Rules 5.3 and 4.7). These can be either extracted from the drawings and enclosed as part of the tender documentation or specially prepared for inclusion with the written description on the facing page (see figure 7.10). The quantity (nr) is established using the same technique as used to find the number of rafters or joists.

It will also be necessary to measure truss clips which secure the foot of each truss to the wall plate. These should be enumerated in accordance with SMM7 G20.22 (see figure 7.11).

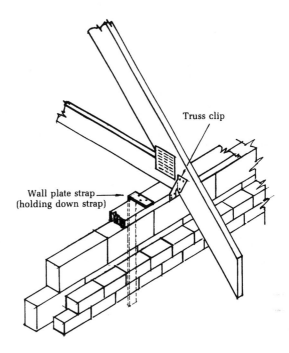

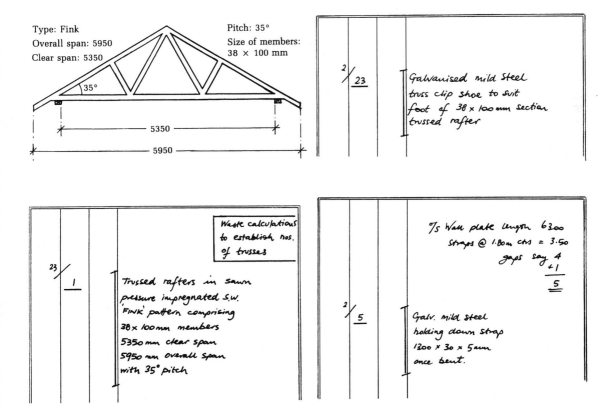

Figure 7.10   Trussed rafters.

Figure 7.11   Trussed rafter clips.

7.5 Cross-sectional sizes of sawn softwood

The standard sawn softwood sizes are given in table 7.6.

Table 7.6 Cross-sectional sizes of sawn softwood

Thickness	Width								
	75	100	125	150	175	200	225	250	300
16	×	×	×	×					
19	×	×	×	×					
22	×	×	×	×					
25	×	×	×	×	×	×	×	×	×
32	×	×	×	×	×	×	×	×	×
36	×	×	×	×					
38	×	×	×	×	×	×	×		
40*	×	×	×	×	×	×	×		
44	×	×	×	×	×	×	×	×	×
50	×	×	×	×	×	×	×	×	×
63		×	×	×	×	×	×		
75		×	×	×	×	×	×	×	×
100		×		×		×		×	×
150				×		×			×
200						×			
250								×	
300									×

*For 40 mm thickness, designers and users should check availability.

The smaller sizes above the line in this table are normally but not exclusively of European origin. The larger sizes below the line are normally but not exclusively of North and South American origin.

Care should be taken when specifying timber to ensure it is from a sustainable source.

7.6 Worked take-off examples

Worked examples of take-offs for a suspended timber floor (take-off sheets: pp. 81–6; drawings: pp. 86–7) and a felt covered flat roof (take-off sheets: pp. 87–93; drawings: p. 94) are shown on the following pages.

A worked example of a take-off for a traditional cut pitched roof is given at the end of chapter 8.

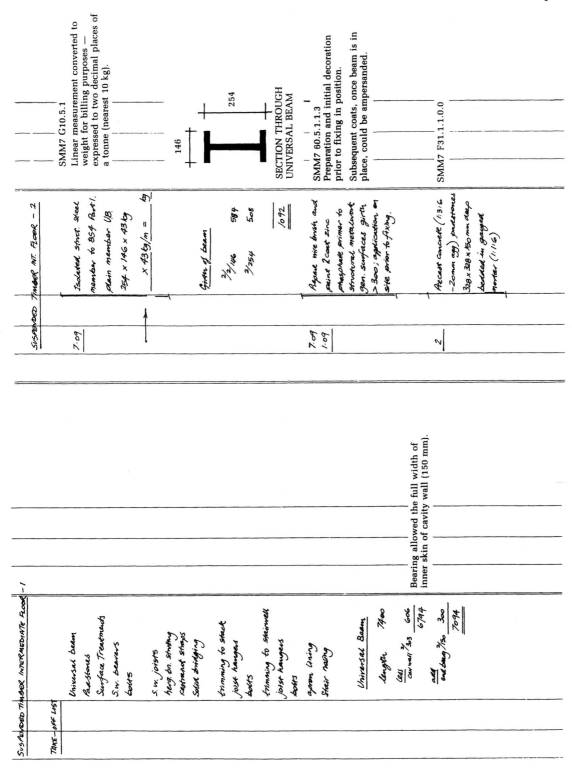

SUSPENDED TIMBER INT. FLOOR - 2

7.09 | Isolated struct. steel member to BS4 Part I. Plain member UB. 254 x 146 x 43kg. 254 x 146 x 43 kg/m = kg

SMM7 G10.5.1
Linear measurement converted to weight for billing purposes — expressed to two decimal places of a tonne (nearest 10 kg).

146   254

SECTION THROUGH UNIVERSAL BEAM

Girth of beam
2/2/146 = 584
2/254 = 508
1092

7.09 | Prepare wire brush and paint 2 coat zinc phosphate primer to structural metalwork. Gen. surfaces girth
1.09 | > 300; application on site prior to fixing.

SMM7 60.5.1.3
Preparation and initial decoration prior to fixing in position.

Subsequent coats, once beam is in place, could be ampersanded.

2 | Precast concrete (1:3:6 —20mm agg) padstones 338 x 328 x 50 mm deep bedded in gauged mortar (1:1:6)

SMM7 F31.1.1.0.0

SUSPENDED TIMBER INTERMEDIATE FLOOR - 1

TAKE-OFF LIST

Universal beam
Padstones
Surface Treatments
S.w. bearers
Bolts

S.w. joists
herg bn. strutting
restraint straps
Solid bridging

trimming to stack
Joist hangers
bolts

trimming to stairwell
Joist hangers
bolts

apron lining
Stair nosing

Universal Beam

length 7400
less 2/353 606
            6794
add end bearg 2/150 300
            7094

Bearing allowed the full width of inner skin of cavity wall (150 mm).

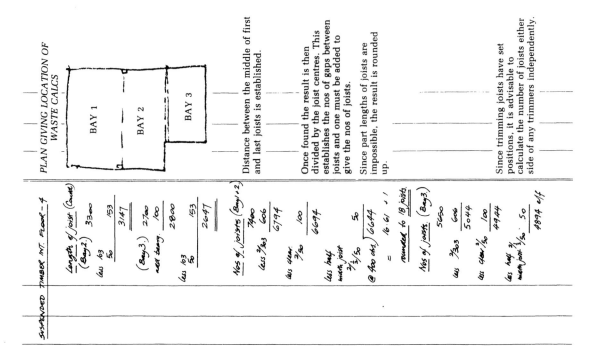

PLAN GIVING LOCATION OF WASTE CALCS

Distance between the middle of first and last joists is established.

Once found the result is then divided by the joist centres. This establishes the nos of gaps between joists and one must be added to give the nos of joists.

Since part lengths of joists are impossible, the result is rounded up.

Since trimming joists have set positions, it is advisable to calculate the number of joists either side of any trimmers independently.

SUSPENDED TIMBER INT. FLOOR – 4

```
Length of joist (Contd)
(Bay 2)          3300
less  153
       50         153
                 3147

(Bay 3)          2700
nett bearing      100
                 2800
less  153
       50         153
                 2647

Nos of joists (Bay 1 + 2)
                 7400
less 2/303        606
                 6794
less clear        100
                 6694
less half
width joist
2/½/50             50
@ 400 chs)  6644 ( 2½/50
           = 16.61 + 1
   rounded to 18 joists

Nos of joists (Bay 3)
                 5650
less 2/303        606
                 5044
less clear 2/50   100
                 4944
less half 
width joist 1/50   50
                 4894  e/f
```

SMM7 G20.8.0.1.0

Length (ref. SMM7 G20.M4)

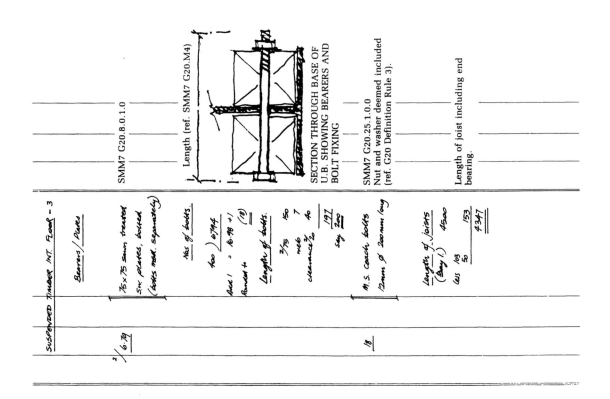

SECTION THROUGH BASE OF U.B. SHOWING BEARERS AND BOLT FIXING

SMM7 G20.25.1.0.0
Nut and washer deemed included (ref. G20 Definition Rule 3).

Length of joist including end bearing.

SUSPENDED TIMBER INT. FLOOR – 3

```
Bearers / Plates
2/6.74

75 x 75 sawn treated
Sw plates, bolted
(bolts meas. separately)

Nos of bolts.
 400 ) 6794
Add 1 = 16.98 +1
Rounded to  (18)

Length of bolts.
 2/75    150
 web       7
 clearance 2/20  40
          197
     say  200

M.S. coach bolts
12mm Ø 200mm long

18

Length of joists
(Bay 1)   4500
less  153
       50   153
          4347
```

SUSPENDED TIMBER INT. FLOOR - 6

Horiz. rest. straps

s/wall length   4500
                3300
                2700
             2/ 10500
              = 21000

less 1st joist
(both sides)
2/3/200   8000
@ 2000 ctrs) 13000
= 6.500 +1
rounded to 8 nos

Straps galv m.s.
vertical restraint
1200 x 30 x 5mm

| 8 |

SMM7 G20.20.1.0.0
Measured where joists run parallel with external walls.

Nogging to Rest. Straps
length 2/400 800
clearance   50
half joist width 25
             875

50 x 150mm sawn
treated s.w. strutting,
strap, 225mm deep
joists

| 8/ | 0.89 |

SMM7 G20.10.2.0.0
Noggings measured as fixing for vertical restraint straps.

SUSPENDED TIMBER INT. FLOOR - 5

Nos of joists (Bay 3) Cont.
@ 400 ctrs) 4894 b/f
= 12.23 +1
rounded to 14 joists

50 x 225 sawn   (Bay 1
s.w. SC grade   (" 2.
treated floor   (" 3
members

18/	4.35
18/	3.15
14/	2.65

SMM7 G20.6.0.1.0

Herringbone Strutting

Bay 1 span a.b. 6794
Bay 2 span   6744
less struts   2950
                3844
Bay 3 span   5060
less   7303
                5044

50x25mm sawn treated
s.w. joist strutting,
herringbone, 225mm
deep joists

	6.79
	3.84
	5.04

SMM7 G20.10.1.1.0
Strutting measured in linear metres over joists (G20 Measurement Rule 1).

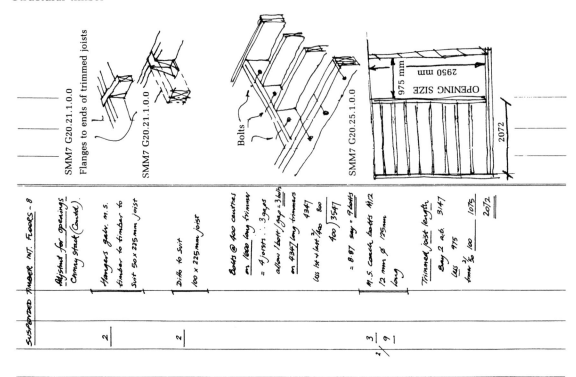

SMM7 G20.21.1.0.0
Flanges to ends of trimmed joists

SMM7 G20.21.1.0.0

Bolts

SMM7 G20.25.1.0.0

OPENING SIZE 2950 mm

975 mm

2072

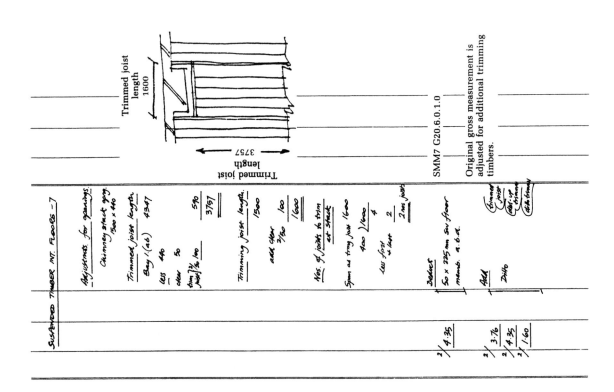

Trimmed joist length 1600

Trimmed joist length 3757

SMM7 G20.6.0.1.0

Original gross measurement is adjusted for additional trimming timbers.

SUSPENDED TIMBER INT. FLOOR - 10.

Adjt for stairwell opg (contd)

7	Hangers galv m.s. timber to timber to suit 50 x 225mm joists	SMM7 G20.21.1.0.0
1	Ditto to suit 50 x 225mm joist	SMM7 G20.21.1.0.0
17	Bolts @ 400 centres tring joists 3147 " " 3100 400) 6247 = 15.6 + 1 rounded to 17 m.s. coach bolts M12 12mm ø 125mm long	SMM7 G20.25.1.0.0

Nos of joists required to trim @ stairwell

2950

STAIRWELL

1 2 3 4 5 6 7 8 9

SUSPENDED TIMBER INT. FLOORS - 9

Adjt for opgs -
Stairwell (Contd)

Trimming joist length
2950
end bearing 150
3100

Nos. of joists to trim
@ Stairwell 2950
@ 400 of) 2950
7.375
Excluding 1st and last
∴ rounded to 8 nos

8	3.15	Deduct 50 x 225mm sw floor memb a.b.		SMM7 G20.6.0.1.0
8	2.07	Add 50 x 225mm sw floor memb a.b.	Trimmed joist Extra trying joist	SMM7 G20.6.0.1.0
	3.15			
2	3.10			

Suspended timber intermediate floor (section and details)

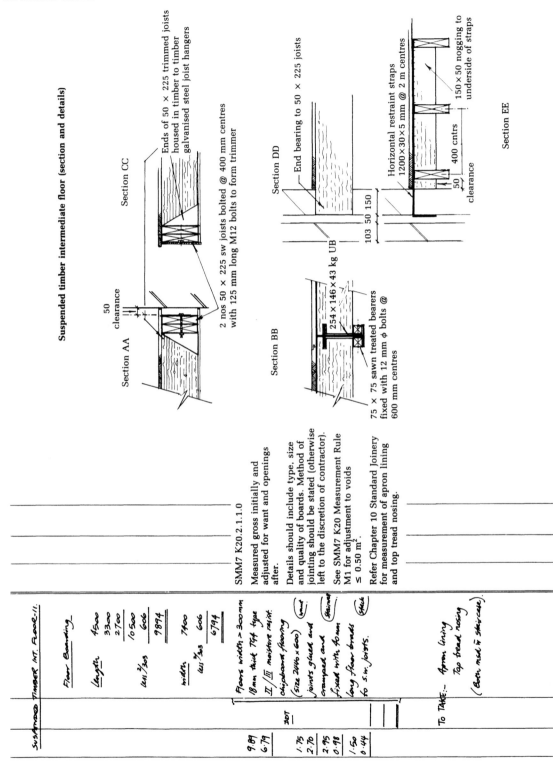

Section AA

Section CC

50 clearance

Ends of 50 × 225 trimmed joists housed in timber to timber galvanised steel joist hangers

2 nos 50 × 225 sw joists bolted @ 400 mm centres with 125 mm long M12 bolts to form trimmer

Section BB

Section DD

End bearing to 50 × 225 joists

254 × 146 × 43 kg UB

103 50 150

75 × 75 sawn treated bearers fixed with 12 mm φ bolts @ 600 mm centres

Section EE

Horizontal restraint straps 1200 × 30 × 5 mm @ 2 m centres

150 × 50 nogging to underside of straps

400 cntrs

50 clearance

SMM7 K20.2.1.1.0

Measured gross initially and adjusted for want and openings after.

Details should include type, size and quality of boards. Method of jointing should be stated (otherwise left to the discretion of contractor).

See SMM7 K20 Measurement Rule M1 for adjustment to voids ≤ 0.50 m².

Refer Chapter 10 Standard Joinery for measurement of apron lining and top tread nosing.

SUSPENDED TIMBER INT. FLOOR II.

Floor Boarding.

Length:-
4500
3300
2700
10500
Less ²⁄₂₀₃   606
        9894

Width:- 7400
Less ²⁄₃₀₃   606
        6794

| | 9.89 |
| 6.79 | |

9894
6794

Floors width > 300 mm (mat)
18 mm thick T&G type
II/III moisture resist
chipboard flooring
(size 2400 × 600)
Joints glued and
cramped and (Skirmet)
fixed with 40 mm
long floor brads (Steel)
to s.w. joists.

1.75	
2.70	
2.95	
0.98	
1.50	
0.44	

To Take:- Apron lining
           Top tread nosing
           (Both mesd in staircase).

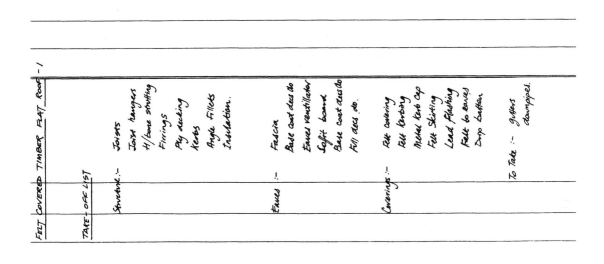

FELT COVERED TIMBER FLAT ROOF - 1

TAKE - OFF LIST

Structure :-
Joists
Joist hangers
H/bone strutting
Firrings
Ply decking
Kerbs
Angle Fillets
Insulation.

Eaves :-
Fascia
Base coat decd do
Eaves ventilator
Soffit board
Base coat decd do
Fill decd do.

Coverings :-
Felt covering
Felt kirbing
Metal kerb cap
Felt Skirting
Lead Flashing
Felt to caves
Drip batten

To Take :- gutters
              downpipes.

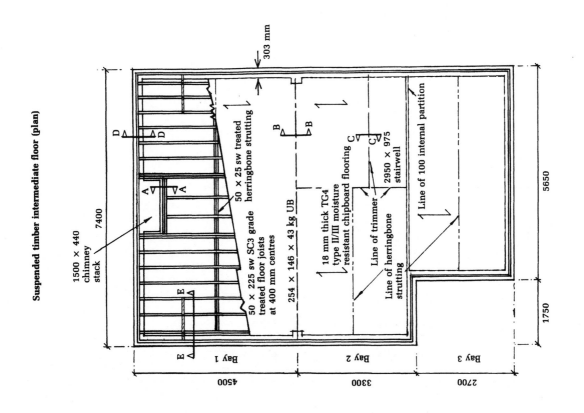

Suspended timber intermediate floor (plan)

FELT COVERED TIMBER FLAT ROOF - 3

Nos. anchor straps

max 2000 ctrs) 8000
gaps = 4 +1 = 5 straps

5	Straps, galv. m.s. anchor, 30 x 5 x 1200mm once twisted, screw fixed to timber joists and brickwork.

SMM7 G20.20.1.0.0

8.00	50 x 50mm sawn treated sw joist strutting, herringbone 225mm deep joists

SMM7 G20.10.1.0.0

Strutting measured over joists.

Firrings length
joist length 4847
add fascia th. 25
4872

Firrings max depth
minimum 13
fall 4872/60 81.2
94.2
max depth say 95mm

Tapering lengths of timber planted to the top of flat roof joists to obtain a fall on the roof decking material.

Two firrings may be cut from a single length of timber.

FELT COVERED TIMBER FLAT ROOF - 2

Roof Structure

nos of joists
length 8000
less
c. joist ½/50   50
jst centres 400) 7950
= 19.80 rounded to 20 gaps
+1 = 21 joists

Length of joist 5000
less Cav wall 303
4697
add eaves proj. 150
4847

Waste calculation to establish length and number of roof timbers.

21	4.85	50 x 225mm sw treated stress graded roof members, flat.

SMM7 G20.9.1.0.0

21	Hangers galv. m.s. timber to brickwork to suit 50 x 225mm joists

SMM7 G20.21.1.0.0

Building in is deemed included SMM7 G20 Coverage Rule C3.

FELT COVERED TIMBER FLAT ROOF – 5

	Kerbs	
	max depth 150 mm	Kerbs fall in opposite direction to firrings and thereby provide a level top to the kerb.
	less fall in roof 95 mm	
	$\underline{\quad55\ mm\quad}$	
2/ 4.87	saun sw. treated individual support 75mm wide, max 150 mm deep tapering to 55mm deep. (in 2 nr)	Same dimension as ply deck width.

SMM7 G20.13.0.1.5 |

Angle fillet / Return to kerb / Kerb / Fascia / 75 mm / 75 mm

	Length of L fillets	
	length 8000	
	less kerb th 2/75 150	
	$\underline{\quad7850}$	
	width 2/4872 9744	
	$\underline{17594}$	
	add return @ eaves 2/75 150	
	add facing 2/75 $\underline{\quad150}$	
	$\underline{17894}$	
17.89	saun sw. treated individual support 75 x 75 mm triangular cross section	SMM7 G20.13.0.1.5

FELT COVERED TIMBER FLAT ROOF – 4

	Firrings (Conte).	
2/ 4.87	saun sw. individual supports 50mm wide x 95mm max deep tapering to 13mm min depth (21 nr.)	SMM7 G20.13.0.1.5
Number not required by SMM7 but given to assist estimating the price of materials/cutting.
	Ply deck width 5000	
	less cav wall $\underline{\quad303}$	
	4697	
	add	
	eaves proj. 150	
	fascia md. 25 $\underline{175}$	
	$\underline{4872}$	
8.00 4.87	Roof boarding width >300mm, 25mm thick ext. quality plywood nail fixed to s.w.	SMM7 K20.4.1.1.0

In this instance no need to state 'sloping' (SMM7 K20 Definition Rule D6).

State nature of background (SMM7 K20 Supplementary Info S3). |

FELT COVERED TIMBER FLAT ROOF -7

Fascia (contd)

exposed girth   250
25
275

8.00 | 2 coat stain to wrot sw gen. isolated surfaced girth ≤ 300mm

SMM7 M60.1.0.2.0

8

Grianaile soffit ventilator unit ref: SV609 screw fixed to timber

Notwithstanding SMM7 G20.28.1.0.0 ventilator unit measured linear.

eaves soffit board
width   150
add bearing
3/103   52
202
416

eaves soffit Decoration girth
7/202   404
2/6   12
416

FELT COVERED TIMBER FLAT ROOF -6

Insulation

Length   8000   606
7394
less 3/303

width   5000   606
less 3/303   4394

7.39
4.39 | 150mm thick fibre-glass insulation quilt, between members @ 400mm centres, horizontal.

SMM7 P10.1.3.1.0

Fascia

Base coat stain girth
3/250   750
3/35   50
550   (Front)

8.00 | Wrot sw fascia board width ≤ 300mm 25 x 250mm

SMM7 G20.15.3.2.0

8

Base coat stain to wrot sw gen. surfaced girth > 300mm appl. on site prior to fixing

x 0.55 =

SMM7 M60.1.0.1.4

FELT COVERED TIMBER FLAT ROOF - 9

Girth individual
boards

2/225   450
2/25   50
           500
          Sides

Base coat stain
to wrot s.w. gen.
surfaces girth >
300mm, applic. on
site prior to fixing.

2/2/ 4.72
     0.50

SMM7 M60.1.0.1.0

ISOMETRIC END DETAIL
OF KERB

SMM7 M60.1.0.1.3

END CAP

See SMM7 J41 Coverage Rule C3.
Notwithstanding this; necessary to
measure separate enumerated item
for purpose-made metal end
capping pieces to kerbs.

Exposed girth area
fixed

           500
as under
capping piece 100
           400
          Sides

2 coat stain to
wrot s.w gen
surfaces girth
> 300, irregular
surface.

2/ 4.72
   0.40

Note:- exposed ends
of kerb covered with
purpose-made capping
pieces.

FELT COVERED TIMBER FLAT ROOF - 8

Eaves or verge soft
boards width < 300mm
ext quality plywood
6 x 200mm

8.00

SMM7 G20.16.3.2.0

Base coat stain to
plywood gen. surfaces
girth > 300mm applic.
on site prior to fixing
              m²
     x 0.42 =

SMM7 M60.1.0.1.4

Fascia. (shiplap)
length 5000
less arr nail   303
                4697
add projn to
front fascia   25
                4722
                Sides

2/ 4.72
   0.50

SMM7 G20.15.1.1.0

wrot sw fascia boards
width > 300mm
comprising 25 x 225mm
shiplap boarding
secret nail fixed to
timber.

FELT COVERED TIMBER FLAT ROOF -11

SECTION THROUGH KERB/
∠ FILLET

75

150

75

75

Kerb @ sides

Kerb ht   150
     ht.   75
part ht 150/2 300
       2/150
                7/150
Less ∠ fillet 75
               75
              225
Less ∠ fillet 75
              334 girth

add
∠fillet = √75² + 75² = 106
              334 girth

2/ 4.87

3 layer felt coverings
all a.b.d. coverings
to kerbs girth ≤ 200mm
ex 200 n ex 400mm
dressed over ∠ fillet
one side and over
skirtslap boarding
other side.

SMM7 J41.14.2.0.0

8

Preformed metal kerb
edge capping trim
Autun ref XYZ screw
fixed to sw @ 450mm
centres with counter-
sunk aluminium
screws.

SMM7 J41.19.0.0.0

SMM7 Coverage Rule (J41) C3
deemed to include ends, angles and
intersections. Notwithstanding this
it may possibly be necessary to
measure a separate enumerated item
for metal kerb and capping pieces.

FELT COVERED TIMBER FLAT ROOF -10

Waste calculation to establish base
dimensions for flat roof covering.

Felt covering

length   8000
less kerb 75
∠ fillet 75
2/150   300
              7700

width   5000
less cav. wall 303
              4697
add soffit proj. 150
    fascia   25
              4872

7.70
4.87

Three layer built up
felt flat roof coverings
1st layer Anderson
HT Elastomeric underlay
weighing 1.8kg/m² on
ext. quality ply base
2nd and 3rd layers
HT Elastomeric M/S
weighing 3.8 kg/m²
hot bituman bonded
between layers
finished with white
spar chippings.

SMM7 J41.2.1.0.0

FELT COVERED TIMBER FLAT ROOF – 12

Skirting

length	8000
less kerb 2/75	150
	7850
girth	225
less ∠ fillet	75
	150
add slope $\sqrt{75^2+75^2}$ =	106
Girth =	256 mm

NB Check Masonry Work measurement for inclusion of cavity tray and weephole measurement.

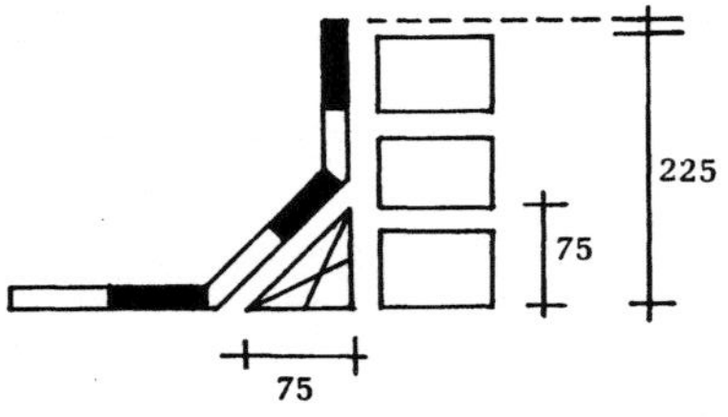

7.85		3 layer built up felt all a.b.d. to skirtings girth ≤ 2.00m ex 200 n.ex 400mm dressed over ∠ fillet one side	SMM7 J41.10.2.0.0

Lead flashing

length	7850
returns 2/150	300
	8150
girth	225
add tuck into bk	50
	275 mm

8.15		Lead sheet flashing Code 4 to BS 1178 275mm girth horizontal wedged into joint of bwk with lead pieces dressing over felt skirting lapped min 150mm at passings.	SMM7 H71.10.1.0.1

FELT COVERED TIMBER FLAT ROOF – 13

Eaves

girth batten	25
	38
o'lap into gutter	25
	88mm

8.00		3 layer felt abd eaves, girth ≤2.00m n.ex 200mm fixed with galvanised clout headed nails to s.w. batten and lapped into gutter (gutters measured elsewhere)	SMM7 J41.4.2.0.0
		&	
		50 x 38mm sawn treated s.w. individual support.	SMM7 G20.13.0.1.0

To take:- gutters
downpipes.

Built up felt timber flat roof (plan)

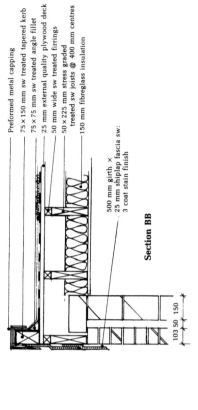

Section BB

Preformed metal capping
75 × 150 mm sw treated tapered kerb
75 × 75 mm sw treated angle fillet
25 mm external quality plywood deck
50 mm wide sw treated firrings
50 × 225 mm stress graded treated sw joists @ 400 mm centres
150 mm fibreglass insulation

500 mm girth × 25 mm shiplap fascia sw: 3 coat stain finish

103 50 150

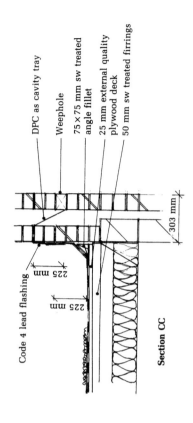

Section CC

DPC as cavity tray
Weephole
75 × 75 mm sw treated angle fillet
25 mm external quality plywood deck
50 mm sw treated firrings

Code 4 lead flashing

225 mm
225 mm
303 mm

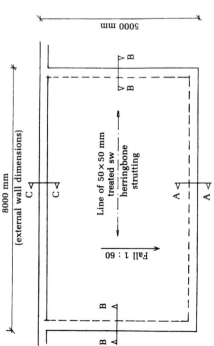

5000 mm

8000 mm
(external wall dimensions)

Line of 50 × 50 mm treated sw herringbone strutting

Fall 1 : 60

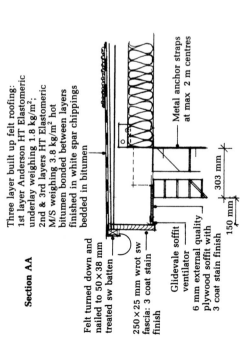

Section AA

Three layer built up felt roofing:
1st layer Anderson HT Elastomeric underlay weighing 1.8 kg/m²;
2nd & 3rd layers HT Elastomeric M/S weighing 3.8 kg/m² hot bitumen bonded between layers finished in white spar chippings bedded in bitumen

Felt turned down and nailed to 50 × 38 mm treated sw batten

250 × 25 mm wrot sw fascia: 3 coat stain finish

Glidevale soffit ventilator
6 mm external quality plywood soffit with 3 coat stain finish

Metal anchor straps at max 2 m centres

303 mm

150 mm

8 Roof coverings

8.1 Introduction

The measurement of pitched and flat roofs conveniently divides between the roof structure and the roof covering. Many of the waste calculations detailed in the previous chapter are equally applicable here. In practice the measurement of roof coverings will follow the measurement of the roof structure and is likely to be carried out by the same person. Eaves and barge boarding together with guttering and downpipes are normally included as part of the measurement of roof coverings.

In previous editions of the Standard Method the trade of Roof Coverings embraced tile, slate and felt roof coverings. The introduction of SMM7 saw tiling, slating and the like being grouped with sheet cladding components such as patent glazing and profiled sheet cladding for both walls and roofs under Section H, Cladding/Coverings. Asphalt and felt-covered roofs are now included in Work Section J, Waterproofing.

The principal unit of measurement for roof coverings is square metres which includes both battens and underlay (SMM7 H60.C1). This makes for an extended but nonetheless necessary description which is best set up on dimension paper as a heading. The descriptive part of the measurement must identify the kind, quality and size of materials, together with the method of fixing. The measurement of the main roof slope area should be followed by adjustments for chimneys and dormers. Alternatively this may be included as part of the measurement of the chimney or dormer. No adjustment is made to the roof covering area for voids of less than one square metre in area.

Ridges, hips, valleys, abutments, eaves, verges and vertical angles are all measured in linear metres and are deemed to include the labour items of cutting, bedding, pointing, forming undercloaks, angles and intersections, and preparing ends. Boundary work to voids is only measured where the void area exceeds one square metre.

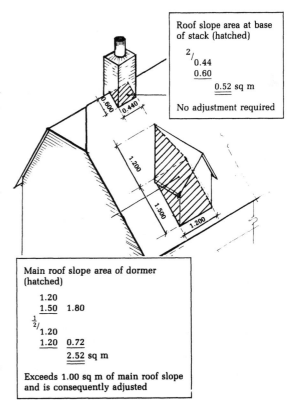

Roof slope area at base of stack (hatched)

2/ 0.44
0.60
0.52 sq m

No adjustment required

Main roof slope area of dormer (hatched)

1.20
1.50   1.80
1/
2/ 1.20
1.20   0.72
2.52 sq m

Exceeds 1.00 sq m of main roof slope and is consequently adjusted

Figure 8.1 Measurement of a dormer window.

8.2 Measurement of pitched roof coverings

The area of roof coverings is unaffected by the inclusion of hipped ends and valleys so long as the roof pitch remains constant. To illustrate this, figure 8.2 shows three alternative roof plans all based on the same dimensions and roof

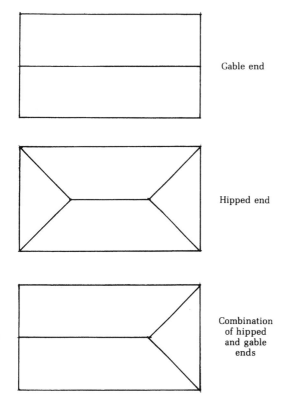

Figure 8.2   Roof plans of pitched roofs.

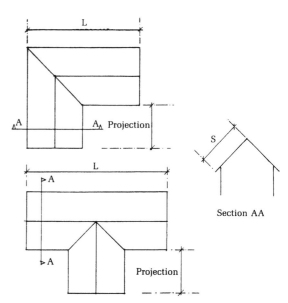

Figure 8.3   Areas of more complex roof plans.

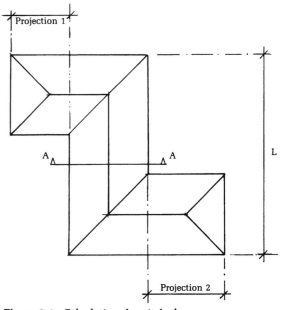

Figure 8.4   Calculating the pitched area.

pitch. When measured, all three will provide the same roof slope area, regardless of whether they are hipped, gabled or a combination of the two.

In each case the sloping roof area can be established by initially ignoring any projections, valleys or hips and simply measuring the main length (L) multiplied by the roof slope (S). Some surveyors choose to enter the dimensions for the roof slope area by recording the plan area of the roof in the dimension column and timesing this by the natural secant of the roof slope in the timesing column. Others prefer to establish the roof slope length as a waste calculation first and then record the plan length and the roof slope length in the dimension column. This is then timesed by two for each roof slope in the timesing column.

For more complicated roof shapes it is advisable to break these down into a main roof length (L) with the projections (P) (see figures 8.3 and 8.4).

In each case the roof covering area can again be established by initially ignoring any projections, hips or valleys and simply measuring the main length (L) multiplied by the roof slope (S) timesed by two for both roof slopes. As a separate operation the plan length of the projection, multiplied by the roof slope (S) again timesed by two, is added to the initial measurement. The squared result is the pitched roof area.

Where the roof has more than one projection the procedure should be repeated for each separate projection.

If the roof slope of the projection differs from the main roof slope the initial measurement should be carried out as before, ignoring any projection. After recording the slope

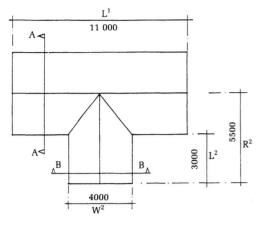

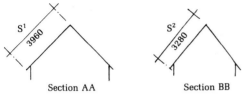

Section AA   Section BB

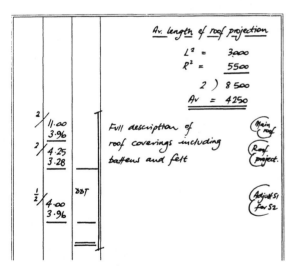

Figure 8.5 The area of a roof which combines different pitches.

area of the projection an appropriate adjustment should be made for the previous over-measurement (see figure 8.5).

8.3 Descriptions for pitched roof coverings

Interlocking concrete tiles are the most commonly used roof coverings since they are relatively cheap to purchase and fix. They are manufactured in a number of different profiles, colours and finishes (see figure 8.6). The large-scale use of the more traditional roofing materials of clay plain

Table 8.1 Common measurement items for most roof coverings

Item	Unit of msmt	SMM7 ref.
Tiles/slates (inc. battens and underlay)	m²	H60.1
Abutments	m	H60.3
Eaves (plain tiling only)	m	H60.4
Verges	m	H60.5
Valleys	m	H60.9
Ridges, hips, vertical angles	m	H60.6/7/8
Fittings	nr	H60.10
Holes	nr	H60.11

tiles and natural slates is limited by the cost of supply and fixing. Both are manufactured in cheaper alternatives (concrete plain tiles, figure 8.7, and fibre cement slates). Even these are still a more expensive roof covering than an interlocking concrete tile. Examples of common measurement items for most roof coverings are given in table 8.1.

Descriptions for pitched roof coverings should include:

- Kind, quality and size of tile
- Pitch of roof
- Minimum extent of side and end laps
- Method of fixing and number of fixings
- Size and type of batten
- Underlay type, laps and method of fixing

8.4 Flat roof coverings

The rules for the measurement of asphalt and built-up felt roof coverings are detailed in SMM7 Work Section J, Waterproofing. Sheet metal roof coverings are included in Work Sections H71 to H74. Measurement of flat roof coverings should follow the same pattern as pitched roof coverings, with the roof structure being measured first, followed by the covering.

8.4.1 Mastic asphalt roofing (SMM7 J21)

Asphalt roofing is measured in terms of the area in contact with the base in square metres. The roof pitch is given in the description and the width must be classified in one of the following categories: not exceeding 150 mm, 150 to 225 mm, 225 to 300 mm and exceeding 300 mm. Skirtings, fascias, aprons, gutter linings and work to kerbs are measured in linear metres and classified in the same width categories. If this width exceeds 300 mm the work is measured in linear metres, stating the girth in the description. Internal angle fillets, fair and rounded edges, drips, arrises and turning asphalt nibs into grooves are all measured in linear metres. Roof ventilators and collars to pipes are enumerated, stating sizes in the description. Edge

Min. pitch — 30°
Max. pitch — 54°
Size of tile — 380 × 229 mm
Covering @ 305 gauge
    = 16.5 tiles/m^2
Special treatment for measure-
ment purposes (measured linear)
• Left hand verge tile
• Eaves ventilation strip
• Requires tilting fillet —
    SMM7 G.20.9

Redland 49 — interlocking tile

Typical description: Roof coverings, 35° pitch, 380 × 230 mm 'Redland 49' interlocking tiles; with 75 mm head lap nailed alternate courses with aluminium alloy nails, to 38 × 25 mm treated softwood battens fixed with galvanised nails, to and including reinforced roofing felt to BS 747 type IF with 75 mm horizontal and 150 mm end laps.

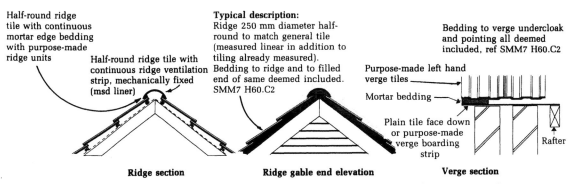

Half-round ridge tile with continuous mortar edge bedding with purpose-made ridge units

Half-round ridge tile with continuous ridge ventilation strip, mechanically fixed (msd liner)

Typical description:
Ridge 250 mm diameter half-round to match general tile (measured linear in addition to tiling already measured).
Bedding to ridge and to filled end of same deemed included.
SMM7 H60.C2

Bedding to verge undercloak and pointing all deemed included, ref SMM7 H60.C2

Purpose-made left hand verge tiles

Mortar bedding

Plain tile face down or purpose-made verge boarding strip

Rafter

Ridge section   **Ridge gable end elevation**   **Verge section**

Figure 8.6 Plain roof tiling – a more attractive and expensive alternative to interlocking concrete tiles.

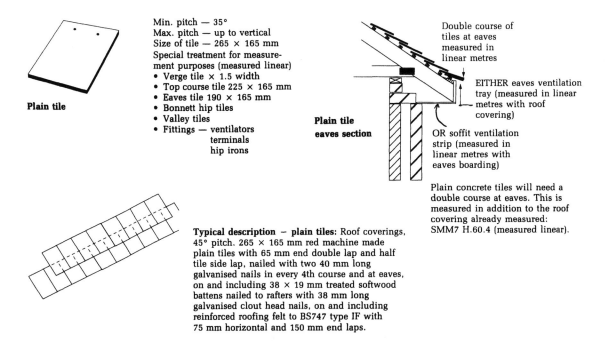

Plain tile

Min. pitch — 35°
Max. pitch — up to vertical
Size of tile — 265 × 165 mm
Special treatment for measure-
ment purposes (measured linear)
• Verge tile × 1.5 width
• Top course tile 225 × 165 mm
• Eaves tile 190 × 165 mm
• Bonnett hip tiles
• Valley tiles
• Fittings — ventilators
                terminals
                hip irons

**Plain tile
eaves section**

Double course of tiles at eaves measured in linear metres

EITHER eaves ventilation tray (measured in linear metres with roof covering)

OR soffit ventilation strip (measured in linear metres with eaves boarding)

Plain concrete tiles will need a double course at eaves. This is measured in addition to the roof covering already measured: SMM7 H.60.4 (measured linear).

Typical description – plain tiles: Roof coverings, 45° pitch. 265 × 165 mm red machine made plain tiles with 65 mm end double lap and half tile side lap, nailed with two 40 mm long galvanised nails in every 4th course and at eaves, on and including 38 × 19 mm treated softwood battens nailed to rafters with 38 mm long galvanised clout head nails, on and including reinforced roofing felt to BS747 type IF with 75 mm horizontal and 150 mm end laps.

Figure 8.7 Plain roof tiling – a more attractive and expensive alternative to interlocking concrete tiles.

trim and preformed angle trim is measured in linear metres and is deemed to include ends, angles and intersections. Adjustments for voids are only made where the void area exceeds one square metre. Boundary work caused by the inclusion of voids is only measured where the void area exceeds one square metre.

8.4.2 Built-up felt roofing (SMM7 J41)

Felt roofing is measured in the same fashion as asphalt roofing, in square metres, giving the area in contact with the base and stating in the description the pitch of the roof. Details of the quality and kind of felt, together with the nature of the base and the method of jointing, must be given in the description. Where the girth of the labour items listed in SMM7 J41.3 to J41.14 is less than 2.00 metres, they should be measured in linear metres giving the girth in 200 mm stages. Where the girth exceeds 2.00 metres, these same labour items should be measured in square metres. Adjustments for voids are only made where the void area exceeds one square metre. Boundary work caused by the inclusion of voids is only measured where the void area exceeds one square metre.

8.4.3 Sheet metal roof coverings

Lead, aluminium, copper and zinc sheet coverings are measured in square metres, stating the pitch of the roof in accordance with SMM7 H71 to H74. The area of covering is adjusted to include an allowance for drips, rolls, seams, welts, laps, upstands and downstands (SMM7 H71−74 Measurement Rule M2). Adjustments for voids are only made where the void area exceeds one square metre. Boundary work caused by the inclusion of voids is only measured where the void area exceeds one square metre.

8.5 Sheet metal flashings/weathering

It is often necessary to measure metal flashings as part of the roof covering to ensure a weathertight construction. Principally this occurs where the slope of the roof is interrupted by chimney stacks or abutting walls. Traditionally the material used for this purpose is lead, which is available in a number of thicknesses referred to by codes (BS1178). Code 4 (1.8 mm) is adequate for most situations, whilst code 5 (2.24 mm) would be used in conditions of severe exposure or for long lengths.

SMM7 clauses H71.10−33 to H76.10−33 inclusive provide the rules for measurement of sheet coverings and flashings. The predominant unit of measurement for flashings is linear metres whilst collars, outlets, saddles and the like are enumerated. In addition to identifying the type of weathering (flashings, aprons, gutters, etc.), the description

Table 8.2   Chimney stack leadwork measurement items

Item of msmt	Unit	Description	SMM7 ref.
Front apron	m	}	H71.11
Back gutter	m	} Dimensioned description	H71.19
Stepped flashing	m	} or dimensioned diagram	H71.*.*.4
Soakers	nr	}	H26.1.*.*.1

should include a dimensioned description or a dimensioned diagram (whichever is the more appropriate) together with a reference to the plane and surface over which the finished component will be dressed. Typically the leadwork to a chimney stack will require the measurement items given in table 8.2 and detailed in figure 8.8.

8.6 Fascia, eaves and verge boarding (G20)

Eaves soffit boarding (fascia and soffit) together with verge (barge) boards not exceeding 300 mm girth are measured in linear metres giving their size in the description. Where boards exceed 300 mm wide they may be measured superficially. Individual or framed supports for eaves soffit boards together with decorations can be measured at the same time.

Eaves to gable-end roofs require boxing at exposed ends (see figure 8.9 spandril end). These are measured as enumerated items stating in the description the kind, quality and finish of the material together with the overall size. Depending on the construction the eaves soffit may include continuous ventilation.

Two separate decoration operations are normally required for all timber surfaces. The first is a priming or sealing application which is applied prior to fixing all joinery timber surfaces. It is normal to knot and stop timber prior to the application of primer and a single description can be written to include all three operations. The second operation is carried out once the timber item has been assembled and combines the separate application of three further coats of paint (two undercoats and one finishing coat) in a single description.

The rules for measuring paintwork distinguish between work under 300 mm wide, which is measured in linear metres, and work in widths exceeding 300 mm, which is measured in square metres. Before dimensions can be recorded it will be necessary to identify which unit of measurement applies and it is therefore necessary to carry out a waste calculation to check the decorating girth. This is illustrated in figure 8.10.

8.7 Rainwater gutters and downpipes

The measurement of rainwater goods completes the work associated with roof coverings. Gutters and downpipes are

Stepped flashing

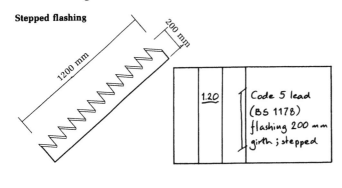

Soaker

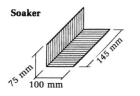

Lead soakers are required as a weatherseal at the exposed bottom edge of the stepped flashing. Typically, these will course with the roof tiling and be tucked below the bottom edge of the flashing. These are enumerated and accompanied by a dimensioned description in accordance with SMM7 H.70.26.

Soakers and similar items are often cut and bent from sheet lead by a plumber and handed to the roofer for fixing. This can be suffixed to the soaker description. In such cases, it is necessary to measure an additional fix only item (SMM7 H60.10.*.*.1).

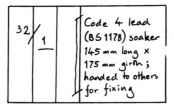

Figure 8.8 Chimney stack leadwork.

Back gutter

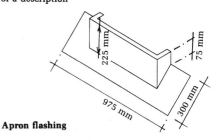

Lead flashing to a chimney stack will include stepped flashings, soakers, back gutters and aprons. The most appropriate way to measure these components is to enumerate and include a dimensioned bill diagram by way of a description

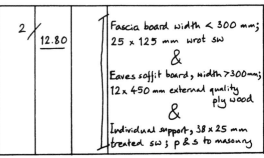

Apron flashing

Boxing to spandril

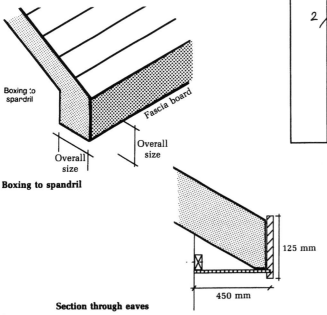

Section through eaves

In addition, it is often convenient to measure the following items at the same time:

M60.1.0.2.4 — Priming before fixing
M60.1.0.1 — Full decoration (based on girth)
G20.18 — Boxing to spandril end
M60.1.0.3 — Decs to ditto (as above)

The measurement of barge boards will follow the same approach as above (included as fascia board SMM7 G20.D10).

Figure 8.9 Fascia, eaves and verge boarding.

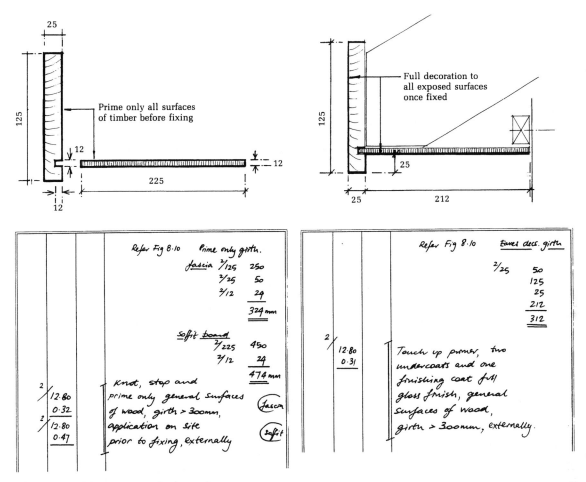

Figure 8.10 Measurement of paintwork.

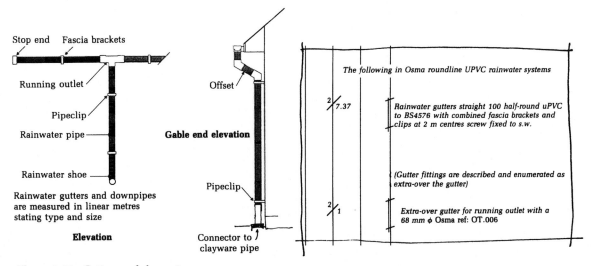

Figure 8.11 Gutters and downpipes.

measured in linear metres over all fittings. The description
should include reference to the type of pipe or gutter, its
nominal size, and the type, method and spacing of fixings;
joints are deemed to be included in the running length.
Fittings, such as running outlets, stopped ends, offsets and
connecting shoes, are enumerated and measured extra over
the pipe or gutter on which they occur. Downpipe fittings
less than 65 mm in diameter must include the number of
connections made to the pipework in accordance with
SMM7 R10.2.3.1 to 4. uPVC rainwater goods are manufactured in different colours and will not normally require any
decoration. Where it is necessary to measure decoration
to pipes and gutters this should be carried out in accordance
with SMM7 M60.8 and M60.9.

8.8 Worked take-off example

A worked example of a take-off for a traditional cut pitched
roof is shown on the following pages (take-off sheets:
pp. 103–12; drawings: p. 113).

Heading can include any detail required by SMM7 G20.S1 to S9.

Halved joint at corners

Halved joint in running length where wall plates exceed 6.00 m in length

SMM7 G20.8.0.1

TRAD. CUT PITCHED ROOF - 2

All structural timber stress graded to CP112 grade S1 group S1

Structural and non structural timber pressure impregnated with preservative

Wall plate dims.
2/ 8900   17800
2/ 2750   5500
ext. wall dims   23300

103 but
Sc. cor. 4/
153
2/2/153   1 224

ext girth w. plt = 22 0%

add
corner plt 4/2/100   800
halved jts 2/50   300
23 176

23.18 ⎡ 100 x 75 mm Sawn
     ⎢ sw plate bedded
     ⎣ in g.m. (1:2:9)

TRADITIONAL CUT PITCHED ROOF-1

TAKE-OFF LIST

Structure :-   Wall plate & straps
Ceiling joists
Rafters
Ridge & ridge batten
Hip & Valley rafters
Purlin and Struts
Binder
Insulation
To take Tank stand / platform.

Covering :-   Tiling battens & felt.
Eaves & eaves vent
Verges
Ridge and ridge vent
Hips
Valleys

Eaves :-   Fascia & true stain
Soffit   "   "   "
Soffit ground
Boxed spandrill ends
Stain finish to eaves

Rainwater
Goods :-   Gutters and fittings
Downpipes and fittings

TRAD. CUT PITCHED ROOF - 4

Nos. ceiling joists - proj

roof proj: 2750
            250
            3000
less c/w wall 288
               2712

add   103
       50
      186

135
 35
-100   o/all clear span 2900
less clear 50
½ joist   25
2/75   150
@ 400 ctrs) 2750
= 6.8 gyps
say 7+1 = 8 joists

Length of joists
main roof + projection 4000
less 1/288   576
                         3424
add
bearing 2/100 200
        3624 mm

50 x 100 sw (joists)
pitched roof
member

SMM7 G20.9.2.1

$\dfrac{22}{8}$/ 3.62

TRAD. CUT PITCHED ROOF - 3

Wall plate straps
w. plate girth 22 o/a
straps @ 1800 ctrs.
= 12.26 gyps
Say 13 + 1 = 14

Galvanised mild steel
holding down strap
900 long x 30 wide x
5mm thick screw
fixed to timber plate
and blockwork.

14

SMM7 G20.20.1.0

Nos. ceiling joists - main
main roof 8900
less
ext wall   288
clearance   50
½ joist   35
b/ends 2/363 726
@ 400 ctrs) 8174
= 20.14 gyps
Say 21 + 1 = 22 joists

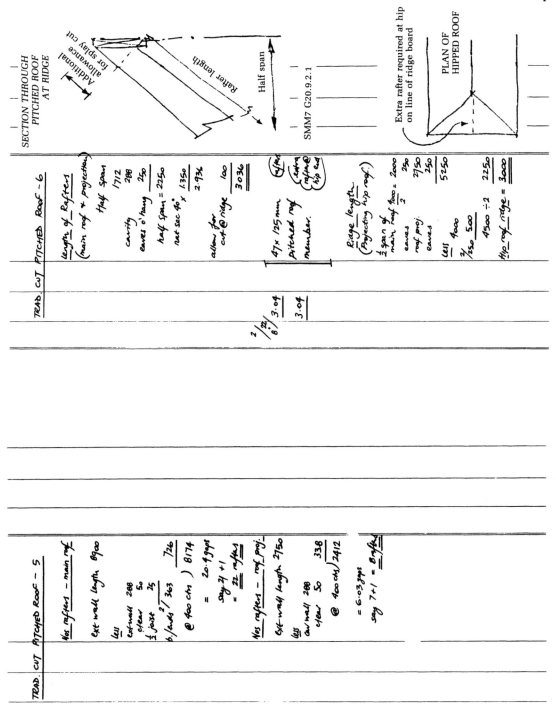

SECTION THROUGH PITCHED ROOF AT RIDGE

Additional allowance for splay cut

Rafter length

Half span

SMM7 G20.9.2.1

Extra rafter required at hip on line of ridge board

PLAN OF HIPPED ROOF

TRAD. CUT PITCHED ROOF - 6

length of Rafters
(main rof + projection)

Half span

          1712
cavity   288
eaves o'hang 250
           2250
half span = 2250
net sec 40° x 1.350
         2.934

allow for
cut @ ridge 100
        3036

47 x 125 mm
pitched roof
member.

        2/22/8/ 3·04

        3·04

(extra rafter @ hip end)

Ridge length
(Projecting hip roof)

½ span of
main roof 4000 = 2000
          2
eaves   250
rof proj.   2750
eaves   250
           5250

Less 4000
2/250 500
     4500 ÷ 2   2250

Hip roof ridge = 3000

TRAD. CUT PITCHED ROOF - 5

Nos rafters - main rof

Ext wall length 8900

Less
ext-wall 288
c/car 50
½ joist 25
b/ends 2/ 363   726

@ 400 ctrs) 8174

    = 20·4 spcs
    say 21 + 1
    = 22 rafters

Nos rafters - rof proj.

Ext-wall length 2750

Less
ext-wall 288
c/car 50   338

@ 400 ctrs) 2412

    = 6·03 spcs
    say 7 + 1 = 8 rafters

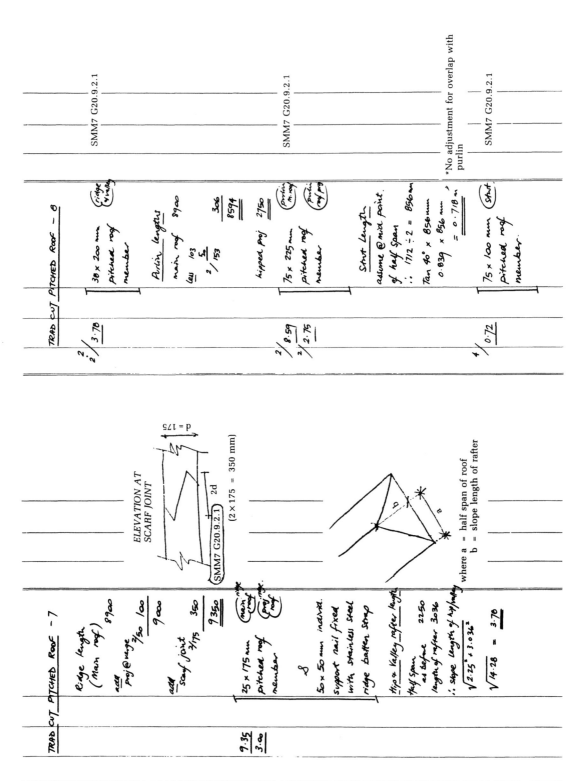

TRAD CUT PITCHED ROOF – 8

2/2/ 3.70

36 × 200 mm
pitched roof
member

(Ridge)
(valley)

Purlin lengths
main roof 8900
less 103
5
2/153

306
8594

hipped proj 2750

(Purlin)
(m. roof)
(Purlin)
(of proj)

2/ 8.59
2/ 2.75

75 × 225 mm
pitched roof
member

Strut Length
assume @ mid point
of half span
∴ 1712 ÷ 2 = 856 mm
Tan 40° × 856 mm
0.839 × 856 mm
= 0.718 m

4/ 0.72

75 × 100 mm
pitched roof
member

(Strut)

SMM7 G20.9.2.1

SMM7 G20.9.2.1

*No adjustment for overlap with
purlin

SMM7 G20.9.2.1

TRAD CUT PITCHED ROOF – 7

9.35
3.00

Ridge length
(Main roof) 8900
add
proj @ verge 2/50 100
9000
add
scarf joint 2/175 350
9350

25 × 175 mm
pitched roof
member

(main roof)
(proj roof)

8

50 × 50 mm indvd.
support nail fixed
with stainless steel
ridge batten strap

Hip & Valley rafter length
Half span
as before 2250
length of rafter 3036
∴ slope length of hip/valley
√2.25² + 3.036²
√14.28 = 3.78

ELEVATION AT
SCARF JOINT

d = 175

2d

(2 × 175 = 350 mm)

SMM7 G20.9.2.1

where a = half span of roof
b = slope length of rafter

TRAD CUT PITCHED ROOF ~ 10

SMM7 P10.2.3.1

Insulation

	length	width
main roof	8900	4000
less cav. wall 2/280	576	576
	8324	3324

proj roof	2750
add eaves	2/yo 288
Cav %	3 288
less Cav %	288
	2000

width (as before) 3324

| 8.32 |
| 3.32 |
| 3.00 |
| 3.32 |

(main roof)

(proj roof)

150mm thick glass fibre quilt laid between joists @ 400 mm ctrs horiz. butt jointed.

TRAD CUT PITCHED ROOF ~ 9

SMM7 G20.9.2.1

Binder length.

(Main roof.)
ceiling joist span (c. to c.) 8174
add
½ joist 25
Scarf jts 100
2/125 250
  8424

(Proj. Hipped roof.)
ceiling joist span (c. to c.) 2750
add ½ joist 2/25 50
each eq 2800

| 2/ | 8.42 |
| 2/ | 2.80 |

(binder)

38 x 100 mm pitched roof member

TRAD CUT PITCHED ROOF - 12

2/ 9.00	Eaves & eaves vent
2/ 3.00	165 x 200 mm eaves tile abd each tile twice nail fixed.
	⟋ (diagonal)
	Eaves ventilation unit with fascia grille and integral apron nail fixed to feet of rafters @ 400mm centres.

SMM7 H60.4.0.0

*NB   Charge to Msmt Rules H60.10.1.1.0 = nr
Since this is impractical the unit of measurement has been changed to linear metres. (Note to be recorded in Preliminaries)

Verges

2/2/ 3.06	Verges including 6 x 150mm non. asbestos board undercloak bedded and pointed in coloured c.m. (1:3) each tile twice nailed with 12g. aluminium alloy nails.

SMM7 H60.5.0.0

TRAD CUT PITCHED ROOF - 11

Roof coverings

```
                    length
main roof            8900
+ verge o'hang   75   100
                     7000

Proj roof           2750
+ eaves o'hang  250   3000

Main/proj roof slope length
rafter length   a.b.  3036

add eaves tile proj. into gutter   25
                                  3061
```

2/ 9.00 / 3.06	165 x 268 mm Redland brown (02) plain roof tiles as roof covering to 40 pitch, laid to a 65mm end lap and half tile side lap, tiles twice nail fixed every 4th course with 12g. alumin alloy nails on and inc. 32x19mm treated sw battens fixed at 100mm gauge with 40mm long galv. nails on and inc. reinfd. roofing felt to BS747 type 1F fixed with galv. clout headed nails with 150mm side and end laps.
2/ 3.00 / 3.06	

SMM7 H60.1.1.0

All embracing description to include S1 details (deemed to include underlay and battens).

Projection of tiles into gutter.

3061
25

TRAD CUT PITCHED ROOF - 14

$^2/$	3.81	Bonnet hip tiles to match gen. tiling each tile nailed with 65mm alumin alloy nails bedded and pointed in coloured c.m. (1:3)

SMM7 H60.7.0.0

&

Purpose made valley tiles to match and course with gen. tiling.

SMM7 H60.9.0.0

9000 x^1

250   2750   250 x^2 x^2 x^2

Eaves boarding -

fascia

(main roof)   8900
add verges $^2/50$ 100
                    9000

(proj hip roof)   2750
eaves proj $^2/250$ 500
                    3250

fascia girth (decs)
$^2/150$ 300
$^2/25$ 50
$^2/12$ 24
           374

TRAD CUT PITCHED ROOF - 13

Ridge

9.00	Ridge, half round pre bed for mechanical fixing with 2 nos. annular shanked stainless steel nails on and including Redland dry vent continuous ridge vent system all fixed in accord. with man. instructions
3.00	

SMM7 H60.6.0.0

Hips/Valleys -

(slope lengths)
½ span roof (ab) 2250
add
eaves tile proj. into gutter   25
                              2275

Slope tiled length as before 3061

$$Length = \sqrt{\tfrac{1}{2}\,span^2 + slope^2\; length}$$

$$= \sqrt{2.275^2 + 3.061^2}$$

$$= 3.814 \; slope \; length \; hip/valley$$

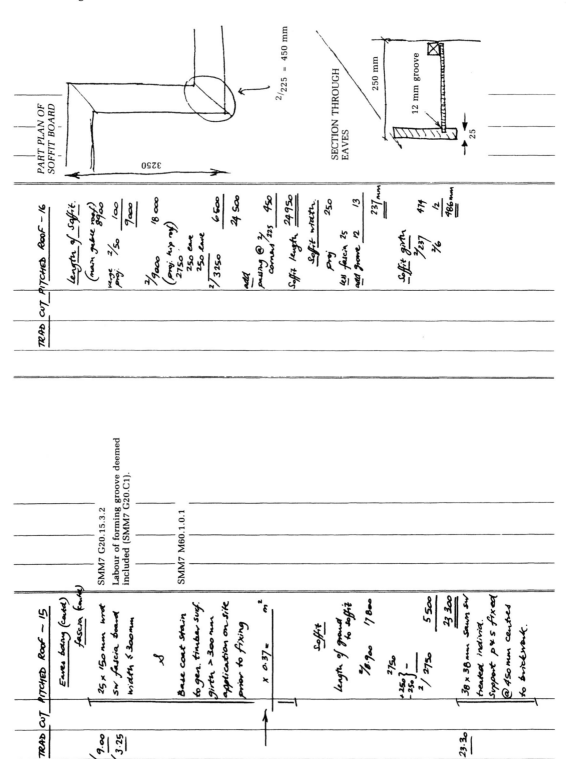

PART PLAN OF SOFFIT BOARD

3250

2/225 = 450 mm

SECTION THROUGH EAVES

250 mm

12 mm groove

25

TRAD CUT PITCHED ROOF - 16

Length of soffit.
(main gable roof)   8900
verge proj   2/50   100
                    9000
2/9000   18 000
(proj. hip rof)
2750
250 eave
250 eave
2/3250   6500
         24 500
add
purlin @ 2/ corners /225   450
                          24 950
Soffit length   24 950

Soffit width.
proj   250
less fascia   25
add groove   12   13
                  237 mm

Soffit girth
2/237   474
½
486 mm

TRAD CUT PITCHED ROOF - 15

Eaves boxing (cont.)
fascia (cont.)

25 x 150 mm wrot sw fascia board width ≤ 300mm

2/9.00
2/3.25

Base coat stain to gen. timber surf. girth > 300 mm application on site prior to fixing

x 0.37 =   m²

SMM7 G20.15.3.2
Labour of forming groove deemed included (SMM7 G20.C1).

SMM7 M60.1.0.1

Soffit

Length of ground to soffit
2/8.900   17 800
2750
+250
-250
2/2750   5 500
         23 300

23.30

38 x 38 mm Sawn sw treated indivd. Support p's fixed @ 450mm centred to brickwork.

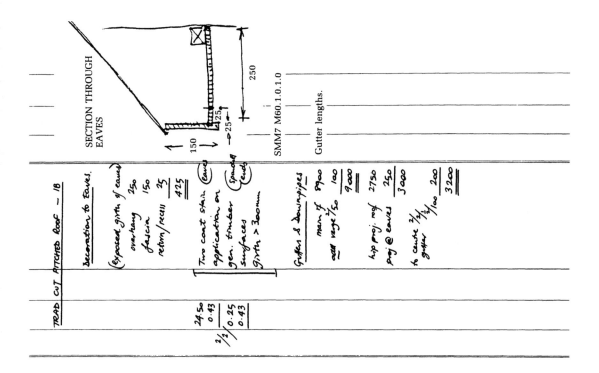

SECTION THROUGH EAVES

250
150
25
→25←

SMM7 M60.1.0.1.0

Gutter lengths.

TRAD CUT PITCHED ROOF – 18

Decoration to Eaves.

(Exposed girth of eaves)
overhang 250
fascia 150
return/roll 25
425

Two coat stain (Eaves) application on (Spandrel gen. timber end) surfaces girth > 300mm

24.50
0.43
0.25
0.43
2/2/

Gutters & downpipes
mean f 9700
add verge 2/150 100
9000

hip proj. roof 2750
proj @ eaves 250
3000

to centre 3/½/100 200
gutter 3200

TRAD CUT PITCHED ROOF – 17.

24.95

6mm thick × 237 wide external quality ply eaves board width < 300 mm

SMM7 G20.16.3.2

Base coat stain gen timber surfaces girth > 300mm, application on site prior to fixing

SMM7 M60.1.0.1.4

× 0.43 = m²

Spandrel End

19mm thick ext. quality plywood spandrel end to eaves overall 250x 450mm

SMM7 G20.18.0.1

2/2

Base coat stain gen timber surfaces isolated area < 0.50 m² applied on site prior to fixing.

SMM7 M60.1.0.3.4

TRAD CUT PITCHED ROOF - 20

2/4.95	65mm Ø uPVC rainwater downpipe to BS 4576 with push fit socket joints fixed with pipe brackets @ 2.00m centres pts to brickwork.
2	E.O. ditto for offset (swanneck) bend 250mm projection
	&
	E.O. ditto for plastic to 110mm clayware adaptor

TRAD CUT PITCHED ROOF - 19

Gutters & Drpipes (cont)

SMM7 R10.10.1.1.1

2/ 9.00 2/ 3.20	100mm Ø uPVC half round rainwater gutters to BS4576 with combined fascia brackets and clips @ 2m centres screw fixed to sw.

SMM7 R10.11.2.1

	Extra over last for:-
2	68mm Ø rw outlet.
4	Ditto:- stopped end
2	Ditto:- external angles
2	Ditto:- internal angles

Assumed waste calcs for height downpipe

$$Gl. \ to \ ff \ level \quad 2400 \ / 150$$

floor joists 225
floor boarding 20   245
                        2795

ff to eaves 2400
              350 2050
(a)   4845

Detail section at eaves

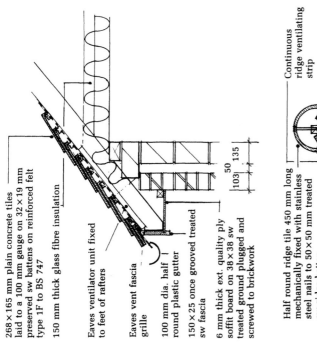

268 × 165 mm plain concrete tiles laid to a 100 mm gauge on 32 × 19 mm preserved sw battens on reinforced felt type 1F to BS 747

150 mm thick glass fibre insulation

Eaves ventilator unit fixed to feet of rafters

Eaves vent fascia grille

100 mm dia. half round plastic gutter

150 × 25 once grooved treated sw fascia

6 mm thick ext. quality ply soffit board on 38 × 38 sw treated ground plugged and screwed to brickwork

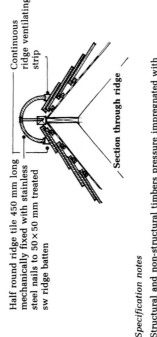

Section through ridge

Continuous ridge ventilating strip

Half round ridge tile 450 mm long mechanically fixed with stainless steel nails to 50 × 50 mm treated sw ridge batten

Specification notes

Structural and non-structural timbers pressure impregnated with preservative.

All structural timber stress graded to CP 112 grade SS group S1.

Continuous ridge and eaves ventilation using dry vent ridge system and plastic eaves ventilation tray all to BS5250.

Base coat and two finishing coats wood stain to eaves.

UPVC gutters and downpipes to BS4576.

Redland plain concrete tiled roof (brown 02) with purpose made valley tiles and bonnet tiled hips.

Roof plan

250

4000

250

2750

250

A

50

50

8900

Struts bearing on loadbearing wall

Line of purlin

Purpose-made valley tiles

38 × 200 valley rafter

Loadbearing partition wall

38 × 200 hip rafter

Bonnet tiled hip

A

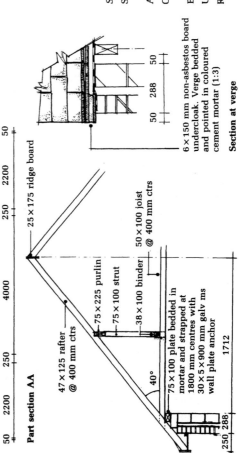

6 × 150 mm non-asbestos board undercloak. Verge bedded and pointed in coloured cement mortar (1:3)

50

288

50

Section at verge

Part section AA

50

2200

250

4000

250

2200

50

25 × 175 ridge board

47 × 125 rafter @ 400 mm ctrs

75 × 225 purlin

75 × 100 strut

38 × 100 binder

50 × 100 joist @ 400 mm ctrs

40°

75 × 100 plate bedded in mortar and strapped at 1800 mm centres with 30 × 5 × 900 mm galv ms wall plate anchor

1712

250 288

103 135

50

9 Internal surface finishes, dry linings, partitioning and suspended ceilings

The term Surface Finishes (SMM7 Work Section M) is used by the Standard Method to identify a number of trades associated with the completion of the floors, walls and ceilings of a building. It embraces several different operations including plastering, screeding, wall and floor tiling, sheet floor finishings, wallpapering and decorating. Associated finishing trades that are excluded from this Work Section but included as part of this chapter are dry lining, proprietary partitioning and suspended ceilings (SMM7 Work Section K).

9.1 General sequence of measurement

As with many other aspects of measurement the key to an efficient and coherent take-off lies in a sensible subdivision of the work and a logical and consistent approach. Since individual circumstance and custom will vary, it is difficult to prescribe a specific approach to measurement. What follows therefore assumes a pattern of finishes which by and large repeat themselves throughout a building. In practice, it is unlikely that a building would be finished in any other way. In any event, the preparation of a schedule will assist in identifying like with like, thereby avoiding a piecemeal or room by room approach.

An example of a finishing schedule accompanies the drawing and take-off at the end of this chapter.

The following sequence of measurement may save a great deal of time by avoiding the need to repeat sets of dimensions:

(1) Floor finishes and screeds
(2) Ceiling finishes and decoration
(3) Skirtings, coving, dadoes and decoration
(4) Wall finishes and decoration

The set of dimensions used to book floor finishes can be linked by way of an ampersand to record ceiling finishes. Similarly, where the internal perimeter of each room is established for the measurement of skirtings, these same base dimensions can be multiplied by the constant floor to ceiling height and used to provide the wall plaster and wall decoration areas. So far as possible descriptions should be grouped in this fashion around a common set of dimensions. It is normal measurement practice to ignore window and door openings, recesses and other features when measuring finishes. In due course an adjustment for this initial over-measurement will be made (see chapter 10).

9.2 General rules of measurement

Work to attached beams is included with ceiling finishes, whilst work to attached columns is included with wall finishes. Finishes to isolated beams and columns must be given separately as should work in staircase areas and plant rooms. Where the floor to ceiling height exceeds 3.50 m (except where caused by staircases) work to ceilings and beams should be so described, stating the height in further 1.50 m stages. The principal unit of measurement for wall, floor and ceiling finishes is square metres. Where the width does not exceed 300 mm the unit of measurement is linear. Individual widths are established on each face or surface. Where this is caused by attached beams and columns it is classified as work to the abutting wall or ceiling. No deduction is made for voids of less than 0.50 square metres or grounds. Since all work is deemed internal the measurer must state in a heading, or as part of the description, whenever work in this section is external.

9.3 Ceiling finishings

The ceiling area is measured in square metres between wall surfaces, stating in the description the thickness and number of coats. Where plasterboard or other sheet baseboard is specified, this should also be included in a single description followed by the thickness and number of skim coats. In cases where the ceiling finishes are identical throughout a floor or storey, it may be appropriate to measure gross, over partitions and internal walls, and subsequently deduct the plan area of internal walls and partitions from the gross ceiling area. (Internal wall dimensions may well be available from earlier measurement.) Care should be taken to identify and measure separately work in staircase areas and plant rooms. Descriptions should include the details required by SMM7 M20.S1 to S8. Where a consistent sequence of measurement is adopted, all the dimensions which relate to each ceiling finish description will be entered as a string of dimensions in the take-off. The need to establish a descriptive heading is therefore eliminated. Careful and consistent signposting is vital in order to avoid confusion at a later stage.

9.4 Floor finishings

Cement and sand screeds together with tiled or wood block floors are measured in square metres, irrespective of their width, and classified by slope in accordance with SMM7 M20.5.1–3 and M40.5.1–3. Sheet flooring, carpet tiling and edged fixed carpeting are likewise classified by slope but measured in square metres only where their width exceeds 300 mm. Work less than 300 mm wide should be measured in linear metres in accordance with SMM7 M50.5.1/2.1–3 and M51.5.1/2.1–3. In some situations it is possible to utilise the previously booked ceiling areas for the measurement of floors. The use of an ampersand to link these two sets of dimensions can save a great deal of time, but care should be taken to ensure the respective plan areas are consistent see figure 9.1).

Floor finishes through door openings will be measured as part of the door opening adjustment as will dividing strips between different types of floor finishing (SMM7 M40.16.4.1.0). Where floor finishes of different thicknesses occur, it is necessary to take up this difference in the screed thickness in order to achieve a level floor finish throughout (see figure 9.2).

9.5 In-situ wall finishings

These normally consist of cement and sand render and plaster. The area in contact with the base is measured in square metres on each face where the width exceeds 300 mm and in linear metres where it is less than 300 mm. The perimeter length of each room should be established as a waste calculation and then transferred to the dimension column where it is followed by the floor to ceiling height. Alternatively, individual room lengths can be recorded in linear metres and this string of linear measurement used in combination with an ampersand to group together the measurement of skirtings, dado rails and picture rails (together with decorations to each), coving, wall finishes and wall decoration. These last two items will require multiplying by the room height (constant dimension) in order to convert them to an area. Signposts must be used to locate the base dimensions to their origin.

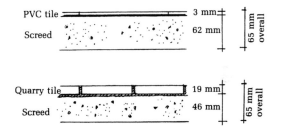

Figure 9.2 Floor finishes of differing thicknesses.

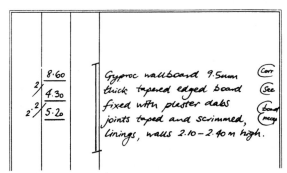

Figure 9.3 Dry lining. Measured in linear metres in accordance with SMM7 K10.

4.80	Plastered ceilings width	(Living)
3.60	>300 mm 12.5mm plstbd	
2.90	fixed with galv. plstbd	(utility)
2.50	nails to s.w. 3mm single	
4.20	coat board finish plaster	(kit)
3.20		
	&	
1.80	65mm thick cement and	(cloak)
1.20	sand screed (1:4) to floors,	
4.00	level and to falls only	(dine)
3.20	≤15° from horizontal	
	wood float finish	

Figure 9.1 Floor finishes. In practice, it is likely that decoration to ceilings and floor finishes could be grouped with these two items.

Internal and external angles together with intersections not exceeding 10 mm radius are deemed included. Rounded angles and intersections in the range 10−100 mm radius are measured in linear metres in accordance with SMM7 M10.16/M20.16. Metal angle beads are measured in linear metres giving a dimensioned description in accordance with SMM7 M10.24.8.1/M20.24.8.1.

Coving, moulding and cornices are measured in linear metres either stating the girth or giving a dimensioned description (SMM7 M10 and M20.17−21). Ends, intersections, internal and external angles are each enumerated and measured extra over (SMM7 M10 and M20.23.1−4).

9.6 Timber skirtings and the like

Timber skirtings, dado and picture rails are measured in linear metres and usually included with the measurement of wall finishings. All are based on room girths and can be ampersanded to the dimensions established for wall plaster purposes (assuming the constant dimension approach is adopted). The labour items of forming ends, angles, mitres and intersections in softwood are deemed included. Similar labour items on hardwood skirtings and the like are only required to be measured where the cross-sectional area of timber exceeds 0.003 square metres (SMM7 P20 Coverage Rule C1). It would be usual to include the decoration of these timber components at this stage and the rules for measurement are included in this chapter (section 9.9).

9.7 Tiled wall finishings

The exposed area on face is measured in square metres where the width exceeds 300 mm and in linear metres where it is less than 300 mm. Work with joints laid out in detail is classified in similar fashion and described as such, keeping it separate from plain tiled work. Internal and external angles not exceeding 10 mm radius, together with intersections, are deemed included. Tiled work exceeding this radius will require special tiles and should be measured in linear metres as extra over the tiling previously measured (SMM7 M40.15.1.1/2 and Definition Rule M40.D8).

9.8 Decorative wallpapers/fabrics

The supply and hanging of decorative wallpaper and fabrics is classified as work to walls and columns or work to ceilings and beams. Where the areas of papering exceed 0.50 square metres they are measured in square metres and where they are less than 0.50 square metres they are enumerated. Where appropriate, lining paper, together with raking and curved cutting, can be suffixed to the description (SMM7 M521−2.1−2.*.1−2). Border strips are measured

in linear metres whilst cutting corners and motifs to a profile are enumerated. It is presumed that the pattern and roll width of the wallpaper should be given as part of the description, although there would appear to be no requirement for the latter to be included under the Measurement Rules of this section. Reference should be made to the Measurement Rules for the adjustment of voids, work in staircase areas and work to ceilings and beams with ceiling heights exceeding 3.50 metres.

9.9 Painting and decorating

The unit of measurement for painting and decorating is based on the girth of the surface receiving decoration. Where this exceeds 300 mm it is measured as an area; where it is an isolated surface and does not exceed 300 mm in girth, it is measured in linear metres; and when it is to isolated areas of less than 0.50 square metres it is enumerated, regardless of its girth. Paintwork to walls, ceilings, beams and columns is classified as work to general surfaces and will appear in the Bill of Quantities as a single conglomerate area (SMM7 M60.1). Paintwork to glazed windows and doors, structural metalwork, radiators, railings, fences and gates, gutters and service pipework should all be classified separately in accordance with SMM7 M60.2−9. All decoration work is deemed internal. When this is not the case it must be stated in the description.

When adopting the group method of measurement it is likely that decoration items will be included at the time of measuring the item being painted. For example, decoration items would normally be ampersanded to the measurement of eaves boarding when measuring roof finishes, to glazed doors when measuring internal doors and to radiators when measuring service installations. As a result of this approach it would be unusual to find a specific section of the take-off for paintwork, and the full extent of the work will only become apparent once it has been billed.

The 'girth of decoration' is an expression used to identify the sectional perimeter of the surface being painted. For example, a timber component of sectional size 175 × 25 mm decorated on all faces (175 + 25 + 175 + 25) would give a girth of 400 mm. The same component fixed to a wall may require decorating only on its exposed face (175 + 25) and give a painted girth of 200 mm. The distinction may seem arbitrary but it is important since it determines the unit of measurement. Paintwork exceeding 300 mm girth is given in square metres, whilst work of less than 300 mm is described as isolated surfaces and measured in linear metres. Because of this classification based on girth, it will be necessary to carry out a waste calculation before measurement can commence. A similar waste calculation may be necessary in order to identify isolated areas of less than 0.50 square metres. For example, a loft hatch size 600 × 900 mm (0.54 m²) would be measured in square metres and classified as a general surface with a girth exceeding 300 mm. Had the loft hatch been 600 × 600 mm (0.36 m²), it would have been classed as an isolated area of less than 0.50 square metres and enumerated. Where this distinction is obvious, for example when measuring decoration to walls and ceilings, no waste calculation is necessary.

9.10 Dry lining

Dry lining provides an alternative wall finish to the more traditional two-coat plasterwork and is measured under SMM7 K10. Work to walls should be measured on the exposed face in linear metres, stating the height in stages of 300 mm, whilst work to ceilings is measured in square metres. Beam and column faces are itemised separately in linear metres, identifying the total girth in stages of 600 mm. In a similar fashion reveals and soffits of openings, together with recesses, are described, stating the width as not exceeding 300 mm or 300 to 600 mm. Linings exceeding 600 mm wide are defined as walls, beams or columns. Work to ceilings in staircase areas and plant rooms should be given separately. Where ceiling heights exceed 3.50 m (except when caused by staircases) they should be so described, stating the height in stages of 1.50 m. Voids not exceeding 0.50 m² to ceilings are ignored. Work to walls is measured on face and is deemed to include all the items listed in SMM7 K10 Coverage Rule C1. No deduction should be made for voids, other than those which extend through full height, full girth or full width (presumably of a sheet). Abutments together with internal and external angles are each measured separately in linear metres. Heavy fittings such as radiators and sanitaryware are enumerated, identifying in the description the type of fitting. Access panels are enumerated and described as extra over the work on which they occur (see figure 9.3).

9.11 Fixed proprietary partitions

As an alternative to a block or studwork partition wall a non-loadbearing proprietary system of plasterboard and cellular core construction might be used (Paramount). Fixed, as opposed to demountable, installations (SMM7 K31) are measured in linear metres stating the height in stages of 300 mm under a very similar set of rules to those used for dry lining. Partitions boarded on one face are described separately to those boarded on both faces. Angles, tee junctions and cross junctions (SMM7 K31.3−5) are all identified and measured in linear metres separately, as are abutments and fair ends (SMM7 K31.6 and K31.8). No deductions are made for voids other than those which extend through the full height or a full panel width (see figure 9.4).

9.12 Demountable partitions (K30)

These should also be measured in linear metres as a mean length, stating the thickness and actual height of the partition in the description (SMM7 K30). A distinction is made in the description between finishes applied on site and those completed at the factory. Trims are also measured in linear metres accompanied by a dimensioned description.

Unlike dry lining and fixed partitioning, openings are enumerated as extra over the partitioning and identified by function and dimensions in the description (see figure 9.5).

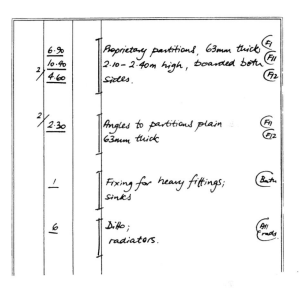

Figure 9.4 Proprietary partitions.

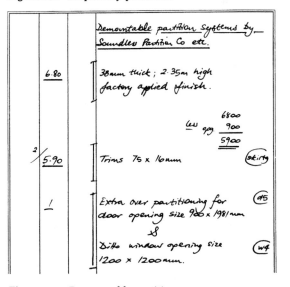

Figure 9.5 Demountable partitions.

9.13 Suspended ceilings

Suspended ceilings are measured in square metres, stating in the description the depth range of suspension together with the thickness of tile and method of fixing to the structure. The Measurement Rules M2 and M4 given in SMM7 K40 are the same as those for dry lining to ceilings. Edge trim would usually be measured separately in linear metres in accordance with SMM7 K40.8 (see figures 9.6 and 9.7).

9.14 Worked take-off example

A worked example of a take-off for internal finishes is shown on the following pages (take-off sheets: pp. 119–24; drawing: p. 124; schedule: p. 125).

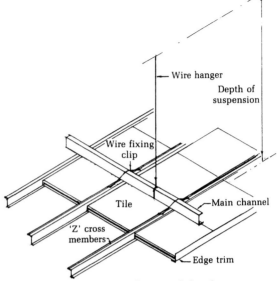

Figure 9.6   Measurement of suspended ceilings.

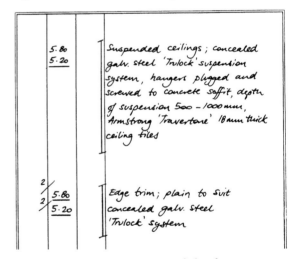

Figure 9.7   Take-off for a suspended ceiling.

INTERNAL FINISHES – 2

Ceilings – C1

Corridor length (08)
2000
2900
1597
partitions 3/100 200
cav wall (ns) 303
7000
less 103 int 153
      50 cav
              6847

Ent. Foyer width (02)
2000
cav wall (int) 303
2303
less 103 int 153
      50 cav
              2150

2·15	
2·60	
2·90	
2·60	
6·85	
2·00	

Plstd. ceilings (02)
width > 300mm (03)
12.5mm plstbd (07)
fixed with galv.
piled nails to
sw, 3mm single
coat board finish
plaster (08)

8

Prepare and apply
basecoat (1nr) and
finishing coats (2nr)
of emulsion paint to
general plaster.
Surfaces girth > 300mm

Waste calculations to establish base dimensions.

SMM7 M20.2.1.2 Scrim to plasterboard joints is deemed included.

NB Noggings may be required to support plasterboard edges at the perimeter and at abutting edges.

SMM7 M60.1.0.1

INTERNAL FINISHES – 1

Take off list:-

Ceilings { Plastered ceilings / Decoration to " / Plasterboard "

Floors { Floor screeds / Decoration/Coatings / Carpets / Quarry tiles

Walls & Skirtg. (Plasterwork to walls / Decoration to " / Skirtings / Decoration to skirting / Render to walls / Quarry tile skirting / Glazed wall tiling / Int L's to gf skirt.

INTERNAL FINISHES — 4

Floors — F1 (Workshop)

12.39	
7.39	

(01) 75mm thick grano-lithic cement and sand screed (1:4) to floors, level and to falls only ≤15° from horizontal power float finish.

SMM7 M10.5.1

Base sealing coat (1nr.) and finishing coats (2nr.) Sadolins floor paint to general surfaces girth > 300 mm

SMM7 M60.1.0.1

Floors — F2 (but Foyer Corridor)

2.00	
2.60	
6.85	
2.00	

(02) 75mm thick c.&s. screed (1:4) to floors level and to falls only ≤15° from horiz. floated finish.

SMM7 M10.5.1

INTERNAL FINISHES — 3

Ceilings — C3 (Male/Female WC)

2/ 1.60	
2.60	

(04/05) Plaster ceilings width > 300mm 12.5mm duplex (foil backed) plstbd. fixed with galvanised plstbd. nails to sw. 3mm single coat board finish plaster

SMM7 M20.2.1.2

Prepare and apply basecoat (1nr) and finishing coats (2nr) of eggshell emulsion paint to general plstd surfaces girth > 300mm

SMM7 M60.1.0.1

Ceilings — C2 (Shop)

2.00	
2.60	

(06) Plastbd. lining 12.5mm thick fixed with galv. plstbd nails to sw ceilings

SMM7 M20.2.1.2
Since this item requires no (wet) finish it might have been included as plasterboard dry lining and measured in accordance with SMM7 K10.2.5.0.

INTERNAL FINISHES – 6

Floors – F4
(Male/female WC)

Screed thickness
o'all   75 mm
g. tile thick   19 mm
C+S screed = 56 mm

SMM7 M10.5.1

56mm thick C+S screed (1:4) [05/06] to floors level + to falls only ≤15° from horizontal

2/ 2.60
   1.60

SMM7 M40.5.1

&

150 x 150 x 19mm heather brown quarry tile to cement and sand base bedded and pointed in c.m. (1:4) to floors level and to falls only ≤15° from horizontal, plain

Floors – F5 [05/06]

SMM7 M10.5.1

75mm thick granolithic cement and sand screed (1:4) to floors, level and to falls only ≤15° from horizontal, floated finish.

2.60
2.00

INTERNAL FINISHES – 5

Floors – F2 (cont'd)

SMM7 M51.5.1.1

Heavy duty carpet to floors width >300mm, perimeter fixed with carpet gripper strip on and including Duralay heavy duty underlay butt joints sewn and heat weld seamed

Notwithstanding SMM7 Coverage Rule C1(d) perimeter fixing has been included in the description.

2/ 2.90
   2.60

Floors-F3 (Reception/Manager) [07/03]

75mm thick C+S screed (1:4) to floors level + to falls only ≤15° from horizontal, floated finish.

SMM7 M10.5.1

&

Medium duty carpet to floors width >300mm perim. fixed with carpet gripper strip on and including Duralay medium duty underlay butt joints sewn and heat weld seamed

SMM7 M51.5.1.1

INTERNAL FINISHES – 8

walls / skirting (contd)
room girths

(02) 2/2591   5197
     2/2150   4300
           9497

(03)(07) 2/2900   5800
       2/2597   5194
             10994

(08) 2/6847   13694
     2/2000   4000
            17694

2/ 9.50
  10.99
  17.69

25 x 150mm sw
bullnosed skirting (03)(07)(06)(08)
pvs fixed to masonry

SMM7 P20.1.1.0

decs girth   8
2/150   300
2/25   50
      350

Priming / sealing coat (1nr) general surf girth > 300mm to s.w. application on site prior to fixing

x 0.35 =   m²

8

SMM7 M60.1.0.1.4

INTERNAL FINISHES – 7

Wall finishes – W1
(01) workshop
(06) Store

room girth (01)
2/12394   24788
2/7394   14788
         39576

room girth (06)
2/2000   4000
2/2597   5194
      9194

39.58
3.90
9.19
2.40

Sealing coat (1nr) and finishing coats (2nr) paint to blockwork general surfaces girth > 300mm

SMM7 M60.1.0.1

Check the take-off for the internal skin of both cavity wall and internal partitions. If not measured fair face make appropriate adjustments.

Wall finishes – W2 and Skirting – S1
(02) Ent. Foyer
(03) Reception
(07) Manager
(08) Corridor

INTERNAL FINISHES - 9

walls/skirtings (Contd)
W2/51

knot, stop and undercoat (2nr), finishing coat (1nr) full gloss paint fin. to general SW isolated surfaces girth <300mm

8

SMM7 M60.1.0.2

Plasterwork to walls width > 300mm, 13mm thick two coat work to blockwork base

× 2.40 =
m²

SMM7 M20.1.1.1

Check plasterwork to widths of less than 300 mm girth caused by door openings and adjust with the door opening as appropriate.

8

emulsion pt height
FI > Clg 2400
Less skirtg 150
2250

Prepare and apply base coat (1nr) and finishing coats (2nr) emulsion paint to plastered general surfaces girth > 300mm

× 2.25 =
m²

SMM7 M60.1.0.1

Adjustment for small area of wall plaster covered by skirting and consequently not decorated.

INTERNAL FINISHES - 10

Wall finishes/skirtings
W3/52

(04) male wc
(05) female wc

room girth
3/2597   5/94
3/1597   3/94
8388

2/8.39

Cement and sand render (1:4) to walls width > 300mm 13mm thick two coat work to blockwork base; wood float fin.

× 2.40 =
m²

SMM7 M20.1.1.1

8

150 × 100 × 19mm heather brown quarry tile skirting 100mm high to cts base jointed & pointed in c.m. (1:4)

SMM7 M40.12.1.0

wall tiling height
FI > Clg 2400
Less gt. skirt 100
2300

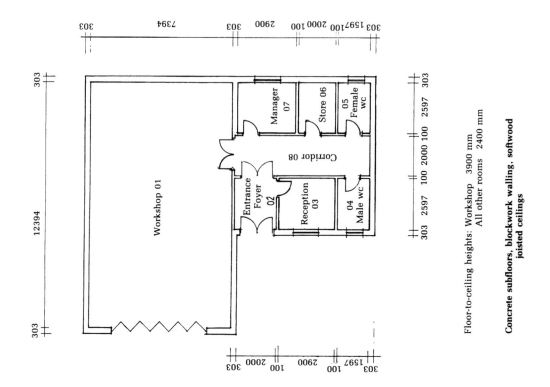

Floor-to-ceiling heights: Workshop 3900 mm
All other rooms 2400 mm

Concrete subfloors, blockwork walling, softwood joisted ceilings

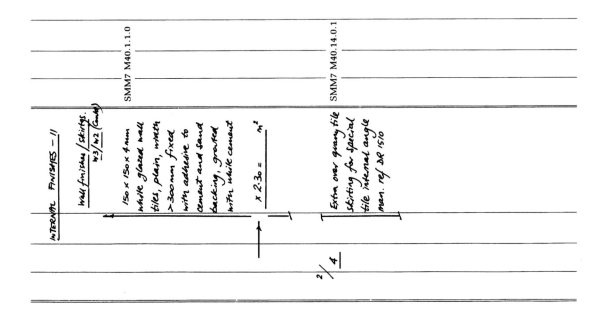

INTERNAL FINISHES - II

Wall finishes/skirtings.
w3/w2 (only)

150 x 150 x 4mm white glazed wall tiles, plain; width >300mm fixed with adhesive to cement and sand backing, grouted with white cement

× 2.30 =   m²

SMM7 M40.1.1.0

Extra over quarry tile skirting for special tile internal angle man. ref QR 1510

SMM7 M40.14.0.1

2/4

Internal finishing schedule

Room	Ceiling	Floor	Wall	Skirting
01 Workshop	C0	F1	W1	S0
02 Entrance foyer	C1	F2	W2	S1
03 Reception	C1	F3	W2	S1
04 Male WC	C3	F4	W3	S2
05 Female WC	C3	F4	W3	S2
06 Store	C2	F5	W1	S0
07 Manager	C1	F3	W2	S1
08 Corridor	C1	F2	W2	S1

Key for finishes:

Ceilings
C0 No finish.
C1 12.5 mm plasterboard with 3 mm skim coat finish. Two coats emulsion.
C2 12.5 mm plasterboard with no finish.
C3 12.5 mm Duplex plasterboard with 3 mm skim coat finish. Two coats eggshell emulsion.

Floors
F1 75 mm thick granolithic screed (1:4) with power float finish. Base sealing coat and two further coats Sadolins floor paint.
F2 75 mm thick cement and sand screed (1:4) floated finish. Heavy duty carpet (edge fixed) on Duralay heavy duty underlay.
F3 75 mm thick cement and sand screed (1:4) floated finish. Medium duty carpet (edge fixed) on Duralay medium duty underlay.
F4 150 × 150 × 19 mm heather brown quarry tile, bedded, jointed and pointed in cement and sand (1:4).
F5 75 mm granolithic cement and sand screed (1:4) floated finish.

Walls
W1 Sealing coat and two coats masonry paint to fair faced blockwork.
W2 Two coat plasterwork 13 mm thick. base coat and two further coats emulsion.
W3 13 mm thick cement and sand render (1:4). 150 × 150 × 4 mm plain white glazed wall tiles (floor-to-ceiling) fixed with adhesive and grouted with white cement.

Skirtings
S0 No skirtings.
S1 150 × 25 mm softwood bullnosed. Knot prime and stop, two undercoats, one finishing coat full gloss finish oil-based paint.
S2 150 × 19 × 100 mm high coved quarry tile skirting with rounded top edge.

10 Standard joinery components

10.1 Introduction

The revisions included as part of the redrafting of the SMM reflect, among other things, the increased dependence on standardisation and prefabrication. Nowhere is this more evident than in the trades traditionally carried out by the carpenter or joiner. This has resulted in a rather disjointed set of rules (with the exception of sections G and L) which appear in odd places under a number of different Work Sections. The segregation of the work in this way will only be evident in the finished Bill of Quantities. For measurement purposes most of the following joinery items would form an integral part of the measurement of a larger component (e.g. architraves with doors, floor boarding with floors, and eaves and verge boarding with roof coverings).

Table 10.1 locates the more usual non-structural timber components using the appropriate SMM7 Work Section reference.

Table 10.1 SMM7 Work Section references to non-structural timber components

Item	Reference
Windows, doors and rooflights	SMM7 L10 and L20
Staircases, balustrading and handrails	SMM7 L30
Skirtings, architraves, window boards	SMM7 P20
Ironmongery	SMM7 P21
Timber flooring	SMM7 K20 Rules grouped with Work Section H, Cladding and Coverings

10.2 Windows, external doors and rooflights (SMM7 L10 and L20)

As with finishes, there is a degree of repetition in the measurement of both windows and doors. In order to maintain a consistent and efficient approach, the preparation of a schedule (figure 10.1) to assist in the measurement of both windows and doors is advised. In the first instance the floor plans should be marked to provide each window with a unique reference. In many cases it is convenient to include external doors at the same time. The same approach would be adopted when measuring internal doors. Whilst this may initially take some time, it will enable like items to be readily recognised and grouped together when dimensions are booked, thereby avoiding unnecessary repetition when taking-off.

10.2.1 General approach to measurement

When dimensions were recorded for the measurement of external walling and wall finishes, no adjustment was made for window and external door openings. A note was made to this effect in the appropriate chapters. This is considered standard measurement procedure and facilitates a speedy and efficient take-off. It is not until this stage that an adjustment is made for this previous deliberate over-measurement. Having first recorded the appropriate items for the window or door, together with any glazing, decoration and ironmongery, the area of walling and finishings are adjusted. This is followed by the measurement of the lintel, closing cavities, damp proof courses, external and internal sills and plasterwork to reveals and soffits. The same approach is adopted for the measurement of rooflights, although the adjustment will only include the omission of roof coverings where the opening area exceeds 1.00 m² (SMM7 H60.M1). Roof members will require trimming and there may be some additional measurement for boundary work (SMM7 H60.M2). Both windows and external doors are most likely to be delivered to site as complete units ready for direct inclusion in the works. Whilst windows will be supplied complete with the frame, external

Location reference	Drawing reference	Manufac- turers' ref.	Window size	Opening size	Glazing	Finish	Additional ironmongery	Lintel			Finish		Sill		Head
								Manufacturer	Catalog ref.	Length	Inside	Outside	Inside	Outside	

Figure 10.1 Schedule headings.

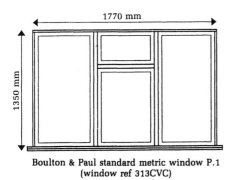

Boulton & Paul standard metric window P.1
(window ref 313CVC)

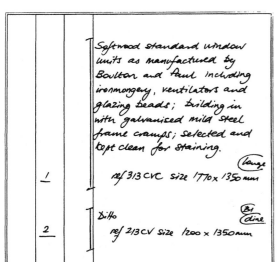

Figure 10.2 Description of prefabricated window units.

doors are likely to require the frame to be provided as a separate item. This has been recognised in SMM7 with the inclusion of a separate measured item for the supply of door frames (SMM7 L20.7). Reference should be made to SMM7 General Rule 9.1 for the measurement of doors and frames supplied as composite items. The measurement is relatively straightforward since an enumerated item, with an appropriate manufacturer's name and reference together with the size, will provide the estimator with sufficient detail for pricing purposes (SMM7 General Rule 6.1). Typically, where prefabricated components are used the measurement of a standard window unit would require a description such as that shown in figure 10.2.

Glazing, painting and pointing frames with mastic will each require separate measurement.

10.2.2 Glazing

Having made an allowance for glazing in rebates, the area of glass is booked. A full description must be given including the kind, quality and thickness of glass, the type of glazing compound, the method of securing the glass and the nature of the frame or surround. Glazing in panes must be classified by pane area in one of two classifications: either less than 0.15 m^2, stating the number of panes in the description, or as panes area 0.15–4.00 m^2.

10.2.3 Painting and decorating

The same approach has been adopted for the measurement of paintwork to glazed windows and doors. The elevation area of the window is recorded measured over the glazed

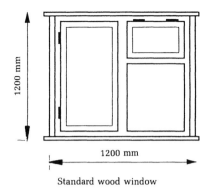

Standard wood window

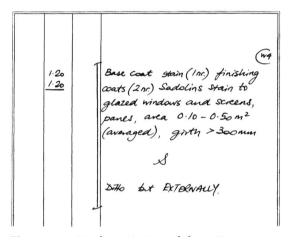

Figure 10.3   Window painting and decorating.

area and the frame, together with any transoms and
mullions. The pane area is classified in one of the four
categories listed in SMM7 M60.2−4.1−4 (less than
0.10 m², 0.10−0.50 m², 0.50−1.00 m² and exceeding
1.00 m²). The paintwork area should be measured to each
face of a window and is deemed to include work to edges
of opening lights. In most circumstances this will exceed
300 mm in girth and be measured in square metres. Work
to glazed doors is classified in similar fashion but should
include an allowance for the edges of glazed doors. Paintwork is deemed internal unless otherwise described (SMM7
M60.D1). Where panes of more than one size occur in a
single window these should be averaged (figure 10.3).

10.2.4 Pointing and bedding frames

If frames are pointed in mastic this must be included as
a separate linear item in accordance with SMM7 L10.9 and
L20.9, stating in the description the pointing compound.
The bedding of frames is included in similar fashion (SMM7

L10.8 and L20.8). Where both operations are required the
two are combined in a single description (SMM7 L10.10
and L20.10).

10.2.5 Ironmongery

Where not supplied as part of the prefabricated component
ironmongery is enumerated stating in the description the
type, manufacturer's reference and the background to which
it is fixed. Most standard windows are supplied complete
with all ironmongery and the measurement of butts, latches,
locks and handles will only be required for doors. Often
a Prime Cost Sum is included for the supply of ironmongery
and where this is the case the measurer need only record
dimensions for fixing. This should be enumerated, stating
in the description the type of ironmongery and the nature
of the base to which it is fixed.

For reference an extract from a typical Bill of Quantities
is included as figure 10.4, showing a number of standard
windows grouped under an all embracing Level 3 description. Both pointing frames with mastic and glazing are
included on the same Bill of Quantity page.

10.3 Internal doors

Standard internal doors, like windows, will be enumerated
and are best described by referring in the description to
the type or style of door, the manufacturer's code and its
dimensions. The general pattern replicates the two-stage
approach adopted for the measurement of windows. Initially, the type of door, its size, any ironmongery and surface
finishes are measured, followed by an adjustment for any
items required to form the opening. Where any number of
internal doors are proposed it is advisable to prepare an
internal door schedule. Door linings are more likely to be
supplied as separate components to the door leaf and in such
cases will require the measurement of the lining as a
separate item. The number of identical door lining sets
should be given in the description and the most convenient
approach is to assume a composite lining set. When this
is not possible, jambs and heads must be given separately
in linear metres, stating in the description the number of
identical lining sets (SMM7 L20.7.1/2.1.1). By and large
most door lining sets are supplied in standard sizes to suit
the more commonly used internal door sizes. Care should
be exercised when recording the door lining width to ensure
the dimension includes the thickness of the plasterwork (or
other finishings) to both sides of the opening, in addition
to the partition thickness (figure 10.5).

The provision of glazing, ironmongery (when not supplied as part of the door leaf) and decorating, should be
measured using the approach outlined above for windows.

					Windows/door/ stairs	
					£	p
	POTTERS CROSS NURSERY					
	PLOT A					
	L WINDOWS/DOORS/STAIRCASES					
	L10 TIMBER WINDOWS/ROOFLIGHTS					
	Softwood stock pattern window units as manufactured by Boulton & Paul Ltd. BP metric p/i/ range including ironmongery; glazing beads and building in with galvanised mild steel frame cramps as work proceeds; selected and kept clean for staining.					
a	1200 × 900 mm ref 209C	1	Nr			
b	1200 × 1200 mm ref 212C	4	Nr			
c	1770 × 1200 mm ref 312C	4	Nr			
d	2339 × 1200 mm ref 412C	2	Nr			
	Pointing frames					
e	mastic compound one side	48	m			
	L40 GENERAL GLAZING					
	Standard plain glass					
f	Glazing panes (6Nr) area less than 0.15 m^2	1	m^2			
g	pane area 0.15–4.00 m^2	16	m^2			
	To collection			£		
	3/42					

Figure 10.4 Typical Bill of Quantities items for window units and glazing.

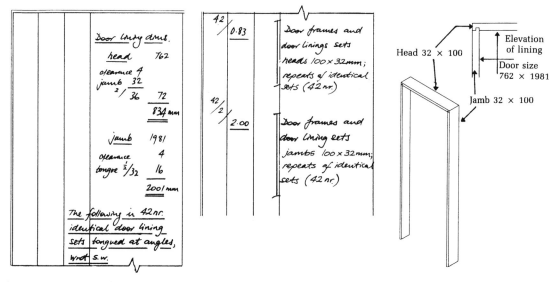

Figure 10.5   Internal doors.

10.4 Timber staircases

As with windows and doors, the majority of timber stair-cases are prefabricated and delivered to site ready to be included in the works. A range of standard staircase components are available from which it is possible to prefabricate a variety of different staircase configurations. Since these are based on standard components and delivered to site prefabricated, they can be conveniently described using the manufacturer's coding system supported by appropriate dimensions. Balusters, newel posts, handrails and the like, where supported by manufacturers' references, are included as part of the staircase description (figure 10.6).

As part of the opening adjustment it may be necessary to adjust the previous measurement of intermediate floor joists and floor boarding. At the very least it will be necessary to check to ensure that the stairwell opening has been formed and if not make the necessary trimming adjustment (see chapter 7).

The surface finishing items of plasterwork and decoration are also likely to require adjustment (see chapter 9).

10.4.1 Purpose-made joinery

Purpose-made windows, doors and staircases require a full description identifying the various components, their respective sizes and the method of jointing or form of construction. Where a component detail drawing is available it is possible to make reference to this in the description, so long as the drawing is included as part of the tender documentation. The latter approach is based on the assumption that component drawings will be available at bill production stage and is by far the most convenient from a measurement point of view. Alternatively, where the design of the component is incomplete, the work should be the subject of a Prime Cost Sum.

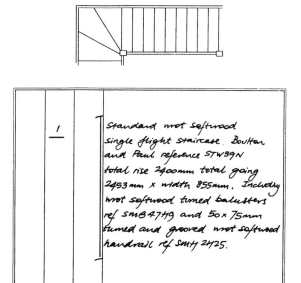

Figure 10.6   Timber staircases.

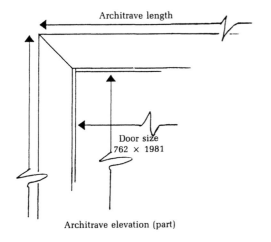

Architrave elevation (part)

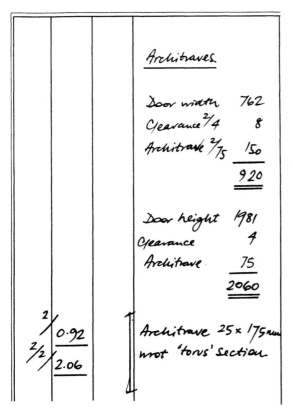

Figure 10.7 Architraves.

10.5 Skirtings, architraves, door stops and window boards (P20)

A number of other isolated joinery components which would normally form part of the measurement of doors and windows are included in SMM7 Work Section P, Building

Fabric Sundries (SMM7 P20, Unframed isolated trims/skirtings/sundry items). The rules for the measurement of architraves, door stops and window boards are included in this section. The unit of measurement is linear metres, stating in the description the dimensioned overall cross-section size.

As with other unframed wrought timber components, most labour items are deemed included (SMM7 P20, Coverage Rule 1). Where labours are carried out on hardwood architraves and window boards and the cross-sectional area exceeds 0.003 m² they should be enumerated and described as extra over the work on which they occur (SMM7 P20.8.1–4). A simple waste calculation would confirm whether the measurement of labours is required (figure 10.8).

10.6 Ironmongery (P21)

Where standard joinery units are supplied complete with ironmongery, no further measurement is required. When not supplied or when additional ironmongery is detailed, each item will be enumerated and described, stating the nature of the base to which it is fixed. Any preparatory work together with screws to match the ironmongery are deemed included. Often it is necessary to include a Prime Cost sum for the supply of ironmongery since a selection may not have been made at the design stage. In these circumstances the fixing of ironmongery must be measured separately. A list identifying ironmongery is included as an appendix to SMM7 (Appendix A, P21).

An internal door would typically require the ironmongery shown in figure 10.9.

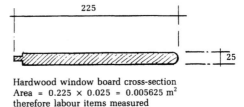

Hardwood window board cross-section
Area = 0.225 × 0.025 = 0.005625 m²
therefore labour items measured

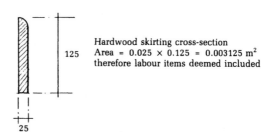

Hardwood skirting cross-section
Area = 0.025 × 0.125 = 0.003125 m²
therefore labour items deemed included

Figure 10.8 The measurement of window and skirting boards.

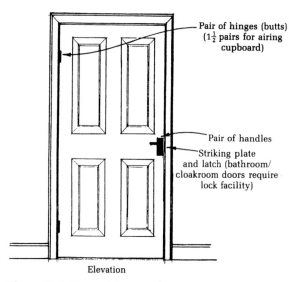

Pair of hinges (butts)
(1½ pairs for airing
cupboard)

Pair of handles

Striking plate
and latch (bathroom/
cloakroom doors require
lock facility)

Elevation

Figure 10.9   Internal doors and ironmongery.

10.7 Worked take-off examples

Worked examples of take-offs for an internal door (take-off sheets: pp. 133–7; drawings: p. 137) and a standard window (take-off sheets: pp. 138–42; drawings: p. 143) are shown on the following pages.

Worked take-off examples   133

INTERNAL DOOR - 1

Take-off list

Door
Ironmongery
Lining
Architrave
Door stop
Decoration to door
   " to door surround

Opening Adjustment
  – Blockwork
  – Plasterwork
  – Emulsion paint
  – Skirting
  – Decoration skirting

Lintel
Floor screed

1/ 35mm thick factory primed hardboard faced moulded and emulsioned internal flush door ref 26C8Y man. by John Carr size 762 x 1981mm

SMM7 L20.1.0.1

Coverage Rule L20.C3 Doors are deemed to include fitting and hanging.

Where doors are supplied with associated frames or linings, measured as a composite item.

INTERNAL DOOR - 2

Supply and fix the following brass ironmongery (BSP reference Vinpack) to hardboard faced standard internal doors :-

2/ 75mm steel washed brass hinges

1/ Pair lever furniture including spindle latch and striking plate

Heading established to prevent repetition in following descriptions.

SMM7 P21.1.1.0

SMM7 P21.1.1.0

Where the supply of ironmongery is included as a Defined Provisional Sum, measurement will be required for the labour of fixing the ironmongery. This is normally measured as 'fix only the following ironmongery', etc.

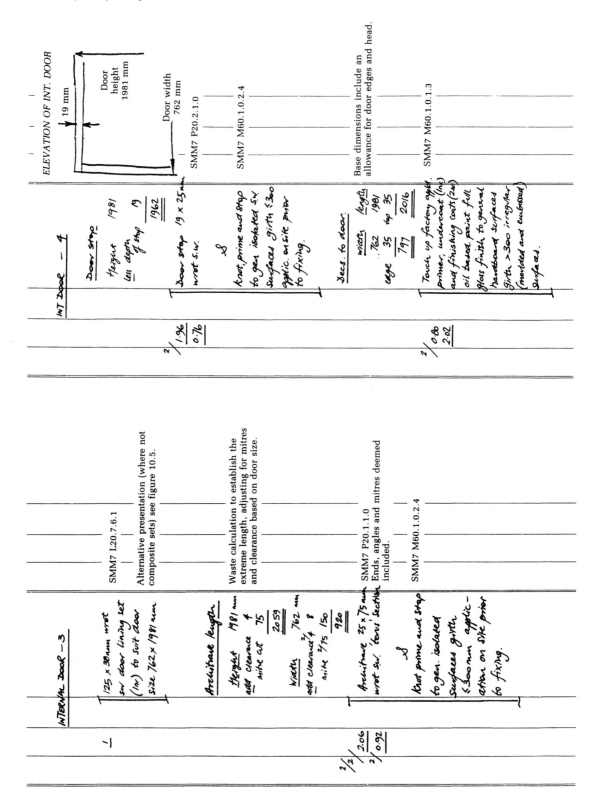

ELEVATION OF INT. DOOR

19 mm

Door height 1981 mm

Door width 762 mm

SMM7 P20.2.1.0

SMM7 M60.1.0.2.4

Base dimensions include an allowance for door edges and head.

SMM7 M60.1.0.1.3

SMM7 L20.7.6.1

Alternative presentation (where not composite sets) see figure 10.5.

Waste calculation to establish the extreme length, adjusting for mitres and clearance based on door size.

SMM7 P20.1.1.0

Ends, angles and mitres deemed included.

SMM7 M60.1.0.2.4

INT DOOR - 6

Opening Adjustment
Height 1981
add lining 38
  2019
Width 762
add lining 2/38 76
  838

Deduct
Lightwt blockwk. walls (all abt.) 103mm vertical

0.84	
2.02	

Deduct
Plstd. walls width >300mm 13mm thick two coat work
×2 =   m²

Deduct
Two coat emuls. paint to gen. plstd surfaces girth >300
×2 =   m²

Blockwork, plasterwork and decoration previously measured through the door opening is now adjusted.

SMM7 F10.1.1.1

SMM7 M20.1.1.1

Base dimensions for blockwork used for plasterwork and decoration adjustment (×2 for each face of walling).

SMM7 M60.1.0.1

INTERNAL DOOR - 5

Decoration, Lengths lining
(w) 762 (l) 1981
2/38 76 76
  838 2019

Girth of lining
2/125 250
2/38 76
  326

2/ 2.02	
0.33	
0.84	
0.33	

Knot prime + stop to gen s.w surfaces girth >300 appln. on site prior to fixg.

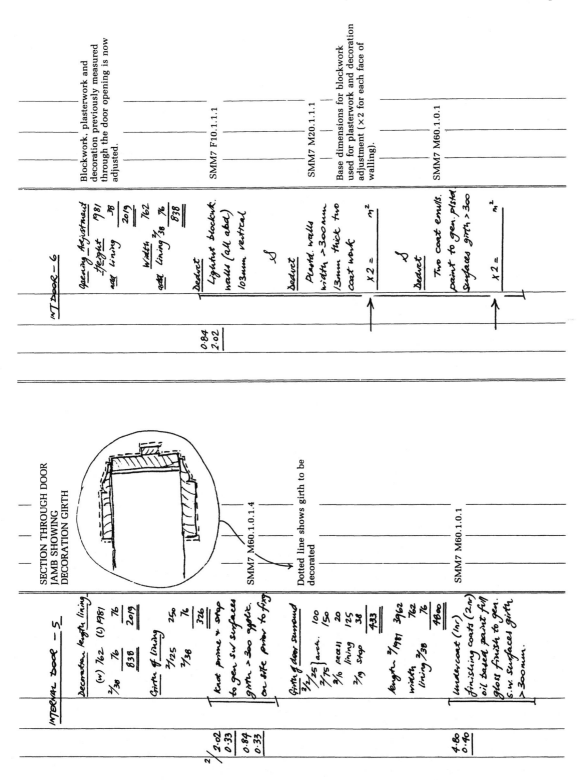

SECTION THROUGH DOOR JAMB SHOWING DECORATION GIRTH

SMM7 M60.1.0.1.4

Dotted line shows girth to be decorated

Girth of door surround
2/2/25 jamb. 100
2/75 150
2/6 reall 20
2/6 lining 125
2/9 stop 38
  433

length 2/1981 3962
width 762
lining 2/38 76
  4800

Undercoat (1nr) finishing coats (2nr) oil based paint full gloss finish to gen. s.w. surfaces girth >300mm.

4.80	
0.40	

SMM7 M60.1.0.1

INT DOOR - 7

Skirting adjustment
door width   762

half lining
38 ÷ 2 = 19
both sides 2/19   38
              150
architrave 2/75   150
              950

Deduct

2/0.95

Skirting 25 × 175 mm
wrot sw 'torus'
section all abd

Deduct

2/0.95   ⊗

Knot prime & stop
to gen. sw surfaces
girth < 300 mm
applic. on site prior
to fixing

  ⊗

Deduct

Undercoat (1nr)
and finishing coats
(2nr) oil based paint
full gloss finish to
general sw surfaces
girth < 300 mm.

Skirting previously measured through the door opening now adjusted.

SECTIONAL PLAN THROUGH DOOR OPG

Length of skirting to be adjusted

SMM7 P20.1.1.0

SMM7 M60.1.0.2.4

SMM7 M60.1.0.2

INT DOOR - 8

2/0.95
  0.18

Two coats emulsion
to gen. plastd.
surfaces girth
> 300 mm   abd

Lintel length

           opening size
door   762
linings 2/38   76
end beng 2/50   300
            1138
nearst manufactured
length size = 1200mm

1

Galvanised steel
Catnic lintel
CN/102, 1200mm
long bedded in
g.m. (1:4)

Door opening width
Blockwork   100
plaster 2/13   26
           126

SMM7 M60.1.0.1

Addition to compensate for emulsion paint previously deducted twice. Once when door opening was adjusted and once when the skirting was measured.

Lintels manufactured in increments of 150 mm.

SMM7 F30.16.1.1

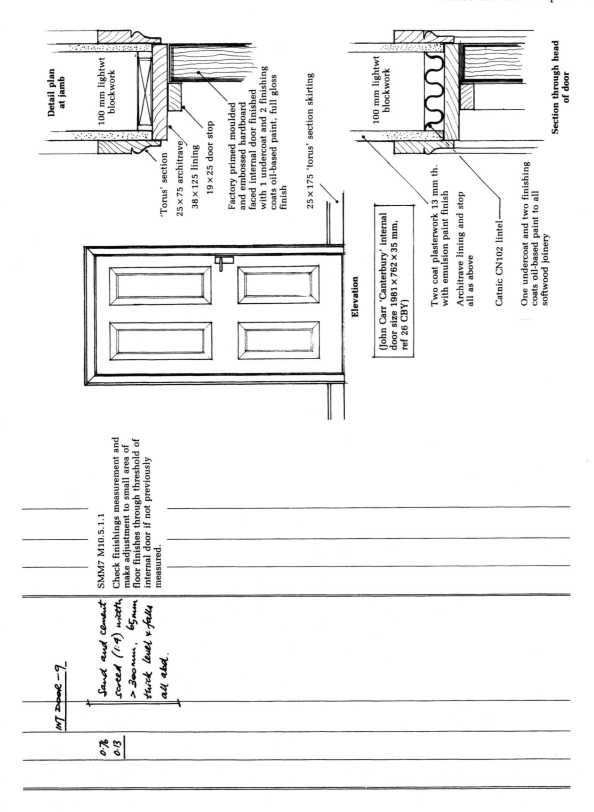

Detail plan at jamb

100 mm lightwt blockwork

'Torus' section

25 × 75 architrave

38 × 125 lining

19 × 25 door stop

Factory primed moulded and embossed hardboard faced internal door finished with 1 undercoat and 2 finishing coats oil-based paint, full gloss finish

25 × 175 'torus' section skirting

Section through head of door

100 mm lightwt blockwork

Two coat plasterwork 13 mm th. with emulsion paint finish

Architrave lining and stop all as above

Catnic CN102 lintel

One undercoat and two finishing coats oil-based paint to all softwood joinery

Elevation

[John Carr 'Canterbury' internal door size 1981 × 762 × 35 mm, ref 26 CBY]

INT DOOR – 9

		Sand and cement screed (1:4) width, 65 mm, > 300 mm, thick level & falls all abd.	SMM7 M10.5.1.1 Check finishings measurement and make adjustment to small area of floor finishes through threshold of internal door if not previously measured.
0.76			
0.13			

SMM7 L10.1.0.1

Manufacturer's Catalogue reference given in lieu of a dimensioned diagram.

Where a number of similar style units are measured, an alternative sequence of description is more appropriate. (Refer figure 10.2)

SMM7 L10.10

STANDARD WINDOW UNIT - 2

1 Standard softwood window unit; Boulton & Paul ref. C312CDC size 1770 x 1200mm including ironmongery ventilators and glazing beads; building in with galvanised mild steel frame cramps; selected and kept clean for staining

—1

Pointing length

2/1770	3540
2/1200	2400
	5940

5.94

[Bed wood frames in gauged mortar (1:4) and point one side in mastic

STANDARD WINDOW UNIT - 1

Take off list :-

Window unit
Bedding and pointing
Glazing
Stain finish

Opening Adjustment
- ext skin cav. wall
- cavity & insulation
- int skin cav wall
- plasterwork
- decoration

Lintel
B.o. edge soldier course
Adjust brickwk for ditto
Cavity tray (dpc)
Closing cavities @ jambs
Plstrwk / Deco - reveals
                - soffits

dpc @ cill
Window board
decoration to wind. bd.

STANDARD WINDOW UNIT - 4

2	1.77
	1.20

Base coat (1nr)
Sikkens Cetol HLS
to s.w. glazed windows
and screens pane
area 0.10 - 0.50 m²
girth >300mm
application on site
prior to fixing.

SMM7 M60.2.2.1.5

Decoration to windows and screens measured in square metres over timber components and glass. Work to opening edges and cutting is deemed included.

1.77
1.20

Finishing coats (2nr)
Sikkens Cetol Filter 7
to s.w. glazed windows
and screens pane area
0.10 - 0.50m² girth >
300mm application on
site prior to fixing

&

Ditto but externally

SMM7 M60.2.2.1

Definition Rule M60.D1 deems work as internal unless otherwise described. This item is assumed internal; the same area is recorded (using an ampersand) for the outer face of the window with a description which includes the word 'external' or 'externally'.

STANDARD WINDOW UNIT - 3

10

Special glass,
sealed double glazing
units comprising 4mm
float (low emissivity
glass (2nr) and 12mm
gap (4:12:4) to
timber frames and
casements with non-
setting mastic and
hard bead pane size
500 x 250 mm;
glazing rebates
20 - 30mm.

SMM7 L40.3.1.2

Double glazing units classified as special glass. SMM7 L40 Definition Rule D4(g).

2

Ditto, 460 x 250 mm;
glazing rebates
20 - 30mm.

SMM7 L40.3.1.2

Pane decl classi. area.
$$m^2$$
$$0.50 \times 0.225 = 0.1125$$
$$0.46 \times 0.225 = 0.1035$$
∴ pane areas 0.10 - 0.50m²

STANDARD WINDOW UNIT – 6

Dimensions	Description	Notes
1	Lintel length / opening 1770 / end beary 2/150 300 / 2070 / nearest manufactured / increment over = 2100mm	Lintel length established by adding the minimum end bearing to the opening size. Prefabricated lintels are manufactured in increments of 150 mm.
1	Galvanised IG steel / insulated lintel / Type L1/S70; 2100 mm / long, bedded in / g.m. (1:4)	SMM7 F30.16.1.1
1·77	Facework (in g.m.– / all abd) ornamental / bands, finish, vertical / brick-on-end course / 215mm wide	SMM7 F10.13.1.1 / Alternatively, the b-o-end course could be measured 'extra-over' in accordance with F10.13.*.*.1 and would consequently need no adjustment.
1·77 / 0·23	Brick courses displaced / by soldier course / 3 x 75mm = 225mm / Deduct / Walls; facework o/s / 103mm thick all / abd.	Deduction only made where full brick courses are displaced. / SMM7 F10.M3 / SMM7 F10.1.2.1

STANDARD WINDOW UNIT – 5

Opening Adjustment

Dimensions	Description	Notes
1·77 / 1·20	Deduct / Walls; facework / one side 103mm thick / abd.	SMM7 F10.1.2.1 The formed opening size and the window dimensions are likely to differ (by a few millimetres).
do	Deduct / Forming cavities in / hollow walls 70mm / wide including rigid / board insulation 40mm / thick all abd.	SMM7 F30.1.1.1 The window sizes given in some manufacturers' catalogues allow for building in; the actual sizes are 5 mm less than those given.
do	Deduct / Walls, insulation / blockwork 150mm / thick abd.	SMM7 F10.2.1
do	Deduct / Plasterwork to walls / width > 300 mm, / 13mm thick 2 coat / work abd	SMM7 M20.1.1.1
do	Deduct / Two coat emulsion / paint to plaster / gen surfaces girth / > 300mm abd.	SMM7 M60.1.0.1

STANDARD WINDOW UNIT - 8

2/ 1:20 / 1:7

Plasterwork to (meas) walls width (Soffit) < 300mm, 13mm thick two coat work.

&

Accessories, metal angle bead for 13mm thick two coat plasterwork

&

Two coats of emulsion paint to plastered general surfaces girth > 300mm

x 0.15 = m^2

Window Board
length 1770
ends 2/75 150
      1920 mm

SMM7 M20.1.2.1   M20 Definition Rule 5 states that work to sides and soffits of openings is classed as work to the abutting walls and ceilings.

SMM7 M20.24.8.1   Alternatively the work to the soffit could be measured/described separately since technically it will, in this case, be to metal lathing base instead of blockwork.

SMM7 M60.1.0.1

STANDARD WINDOW UNIT - 7

Cavity Tray - length
Lintel length 2100
o/hg @ ends 2/50 300
              2400 mm
              width
Brick ht 225
allow for slope
built in ends 2/75 150
                   435 mm

2/40
0.43

Damp proof course (abd) width > 225mm horizontal ad cavity tray bedded in g.m. (1:4)

SMM7 F30.2.2.3.1

Closing cavities 80mm wide with 150mm thick insulation block, vertical.

&

dpc width
block 150
nail fixing 100
             250 mm

SMM7 F10.12.1.1

dpc (abd) width > 225 mm vertical o/edge pinned to back of frame

x 0.25 = m^2

2/ 1:20

SMM7 F30.2.2.1

STANDARD WINDOW UNIT - 10

Area ded. for mastic.

1.92
× 0.225
= 0.432 m²

∴ isolated surface

Undercoat (1nr) and finishing coats (2nr) full gloss finish on primed general isolated area <0.50 m²

SMM7 M60.1.0.3

—

STANDARD WINDOW UNIT - 9

1.92

Medium density Fibreboard (MDF) window board 25 × 225mm rebated and rounded one edge, secret fixed to masonry

&

SMM7 P20.4.1

Rounded and notched ends deemed included.

cill/dpc width
163
86
25
b/i 209

Damp proof course width ≤ 225mm horizontal bedded in g.m. (1:4)

× 0.21 = m²

SMM7 F30.2.1.3

Prime only gen.mr to w.bd.

2/225 450
2/25   50
       500

1.92
0.50

Prime only general fibreboard surfaces girth > 300mm application on site prior to fixing.

SMM7 M60.1.0.1.4

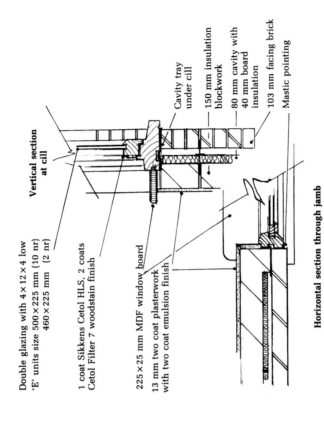

Vertical section at cill

Cavity tray under cill

150 mm insulation blockwork

80 mm cavity with 40 mm board insulation

103 mm facing brick

Mastic pointing

Double glazing with 4 × 12 × 4 low 'E' units size 500 × 225 mm (10 nr) 460 × 225 mm (2 nr)

1 coat Sikkens Cetol HLS, 2 coats Cetol Filter 7 woodstain finish

225 × 25 mm MDF window board

13 mm two coat plasterwork with two coat emulsion finish

Horizontal section through jamb

1200 mm

1770 mm

Elevation
(Boulton & Paul, ref. C312CDC)

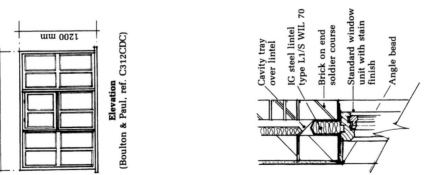

Cavity tray over lintel

IG steel lintel type L1/S WIL 70

Brick on end soldier course

Standard window unit with stain finish

Angle bead

Vertical section at window head

11 Mechanical and electrical service installation

11.1 Introduction

New domestic property will require a piped or cabled supply for the following services:

- Lighting and power
- Hot and cold water
- Heating
- Waste disposal
- Communication
- Safety and security

In non-domestic situations it is likely that a Services Engineer would be employed to provide detailed design and layout drawings which could be used as a base to prepare quantities. So far as domestic property is concerned, such an appointment is unlikely and no such detail will be available. Whilst layout drawings will identify the location of sanitaryware, socket outlets, light switches, boilers, radiators and the like, there will be no indication of pipe and cable runs. These will be left for the plumber or electrician to determine.

Largely because of the specialist nature of service installations, this work is often the subject of a Prime Cost Sum. Whilst the measurement of domestic services is relatively straightforward, there is a presumption that the measurer has a firm understanding of the principles associated with services technology.

Unlike most other Work Sections, the rules for the measurement of mechanical and electrical services and the classification of this work appear in different parts of SMM7. The classification is given in Appendix B and is presented in the following Work Sections:

R   Disposal Systems
S   Piped Supply Systems
T   Mechanical Heating/Cooling/Refrigeration Systems
U   Ventilation/Air Conditioning Systems
V   Electrical Supply/Power/Lighting Systems
W   Communications/Security/Control Systems

The measurement framework appears in Section Y of SMM7 and includes a combination of rules which together provide the basis for recording dimensions in the above Work Sections. Where service installation drawings exist both electrical and plumbing installation for domestic property should be measured in accordance with these rules. Where no such detail exists a Prime Cost Sum may be included in the Bill of Quantities for work to be carried out by a Nominated Subcontractor (SMM7 A51.1). The inclusion of PC Sums is considered in more detail in chapter 4.

This chapter assumes enough drawn detail is available to allow the preparation of dimensions. It commences with the provision of mains services and then describes a typical electrical and plumbing installation to a domestic property. Rather than providing a full take-off for each of these, a take-off list and a set of descriptions based on typical situations has been included.

11.2 Service entry to domestic buildings

The provision of water, gas, electricity and telecom services should be made in a single service trench which terminates in a temporary access chamber near to the building. From this position individual services are ducted to their different meter positions or entry points.

Service trenches are measured in linear metres in accordance with SMM7 P30.1. The description should include the nominal size of the service (not exceeding 200 mm or exceeding 200 mm — in the latter case giving the actual size) together with the average depth of the run to the nearest 250 mm. Earthwork support, backfilling and disposal, together with the other items listed in SMM7 P30.C1, are all deemed included. Separate measured items are required for beds, stop cock pits and any chambers associated with services.

The public utility companies will provide ducts for the contractor to lay in the service trench. At a later stage the

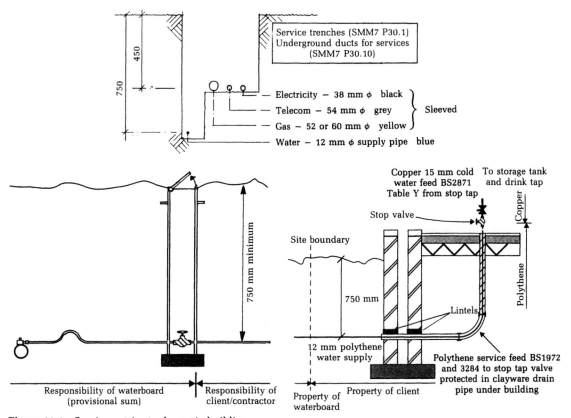

Service trenches (SMM7 P30.1)
Underground ducts for services
(SMM7 P30.10)

Electricity – 38 mm φ black ⎞
Telecom – 54 mm φ grey   ⎬ Sleeved
Gas – 52 or 60 mm φ yellow ⎠
Water – 12 mm φ supply pipe blue

750
450

750 mm minimum

Responsibility of waterboard
(provisional sum)

Responsibility of
client/contractor

Copper 15 mm cold   To storage tank
water feed BS2871   and drink tap
Table Y from stop tap

Stop valve

Site boundary

750 mm

12 mm polythene
water supply

Copper

Polythene

Lintels

Property of
waterboard

Property of client

Polythene service feed BS1972
and 3284 to stop tap valve
protected in clayware drain
pipe under building

Figure 11.1   Service entries to domestic buildings.

services will be sleeved through these ducts and a connection made to the property. These ducts are measured in linear metres in accordance with SMM7 P30.10. (The only exception to this is the water connection which requires no sleeve.) The provision of the service to the point of metering is the responsibility of the appropriate public utility company. A standard charge will be made by the utility company for the connection and payment will normally be required before work can be commenced. The cost of this is usually included in the Bill of Quantities by way of a Provisional Sum based on a quotation from the appropriate company (SMM7 A53.1.2 and Definition Rule D10). The gas and electricity companies will provide meter cabinets together with sleeved rest bends for building in to the cavity walling. An enumerated item should be included for forming and building in meter cabinets giving details in the description of size, fixings, damp proofing, lintel support and any associated service ducting. Where service pipework enters the property an allowance for lintel support to the substructure walling must be made.

In order to make service connections it may also be necessary to include some provision for breaking up and

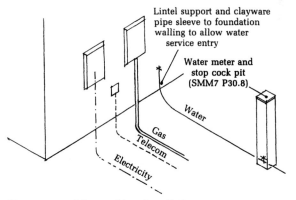

Lintel support and clayware
pipe sleeve to foundation
walling to allow water
service entry

Water meter and
stop cock pit
(SMM7 P30.8)

Water

Gas

Telecom

Electricity

Figure 11.2   Meter cabinet installation.

reinstating the public highway. In the normal course of events this will be carried out by the appropriate utility company and is included as part of their standard charge. At a later stage the completed service installation is tested and, subject to the public utilities approval, a connection is made.

The following text identifies the rules of measurement for the services most commonly encountered in domestic buildings.

11.3 Electrical installation

Before measuring the electrical installation associated with domestic accommodation, it is worth spending some time becoming familiar with both common wiring practice and the graphical symbols which are used to identify electrical components and appliances. The following is offered by way of introduction to the measurement of the electrical installation required in domestic property. For more complex installations reference should be made to the regulations governing electrical installation published by the Institution of Electrical Engineers (IEE).

Electricity can only be transmitted through a conductor when there is a complete circuit from the source, via a conductor, back to the source. Each conductor cable contains a 'live' wire carrying the power to an appliance, a 'neutral' wire carrying the power back to the source and an earth wire which reduces the risk of shock by carrying the current to a circuit breaker or the ground in the event of a short-circuit. The conductor used for domestic supplies is copper wire. In the UK electricity is generated and supplied in three-phase AC (alternating current). For domestic installation only a single phase supply is required and this is delivered to the property via a service cable which terminates in a main fuse mounted in the meter cabinet. From here the connection to the meter is made and the responsibility of the electricity company ends. The consumer unit mounted inside the property steps down the supply amperage by dividing the incoming service into a

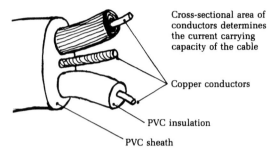

Figure 11.4   PVC-sheathed and insulated twin and earth cables.

Table 11.1   Domestic wiring intallation details

Cable size (mm²)	Wiring circuits	
1.00 or 1.50 2.50	Lighting circuits Power circuits	} One per floor
2.50 10.00 10.00	Immersion heater Cooker Spontaneous shower	} Single circuit per appliance

number of separately fused circuits. Each circuit is protected by its own residual current circuit breaker (RCCB) which detects any abnormal flow or surge of electricity and immediately cuts the supply (figure 11.3).

There are normally three conductors in a cable: a live, a neutral and an earth. The live and neutral are insulated with red and black pvc respectively, whilst the earth is uninsulated. All three wires are then sheathed in an outer layer of insulation. These are consequently known as *pvc-insulated twin and earth cables*. Cables are identified by the cross-sectional area of the conductors and this is expressed in square millimetres (mm²). The larger the area the bigger the current that can safely flow through the cable (figure 11.4). The more common cable sizes and number of wiring circuits found in domestic wiring installations are given in table 11.1.

Unlike most other Work Sections, the rules for the measurement of mechanical and electrical services and the classification of this work appear in different parts of SMM7. As was noted earlier, Appendix B provides the classification for electrical installation under Work Section V, Electrical Supply/Power/Lighting Systems, or Work Section W, Communications/Security/Control Systems. For each of these the measurement framework appears under Work Section Y. The Standard Method distinguishes between cables in two different situations: those which are in final circuits and those which are not. It is important to be able to identify which is which since the principle and approach to measurement is very different.

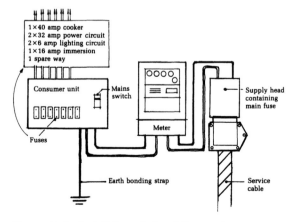

Figure 11.3   Installation of electricity supply.

Work in final circuits   Most installations of a domestic or similar simple nature will be included under this category. Ring mains, lighting circuits, cooker and immersion heater points should all be measured under this category. Each circuit is enumerated (nr) stating in the description the type and size of cabling and the number of socket outlets or lighting points serviced by the circuit (SMM7 Y61.19). As a separate enumerated item socket outlets, light points (luminaires) and other accessories should be enumerated (SMM7 Y74.2 and 5).

Power supply and socket outlets   These circuits are generally arranged as a ring main. A single 2.5 mm^2 twin and earth cable is looped in and out of each socket outlet and is then returned to the same fuse, thereby completing the circuit and forming a ring. Under normal circumstances a separate ring would serve each floor (figure 11.5).

Lighting circuits   Lighting circuits are not installed as rings, although they are still measured as final circuits. Lighting circuits are normally installed with a 'loop-in' arrangement. Cabling runs directly from the fuse board to a ceiling rose which has four terminals. One of these connects to the ceiling rose, one to the switch, one to the lampholder via a flex and one to the next ceiling rose in the circuit. Up to twelve light points can be installed on a single circuit in this way. It is standard wiring practice

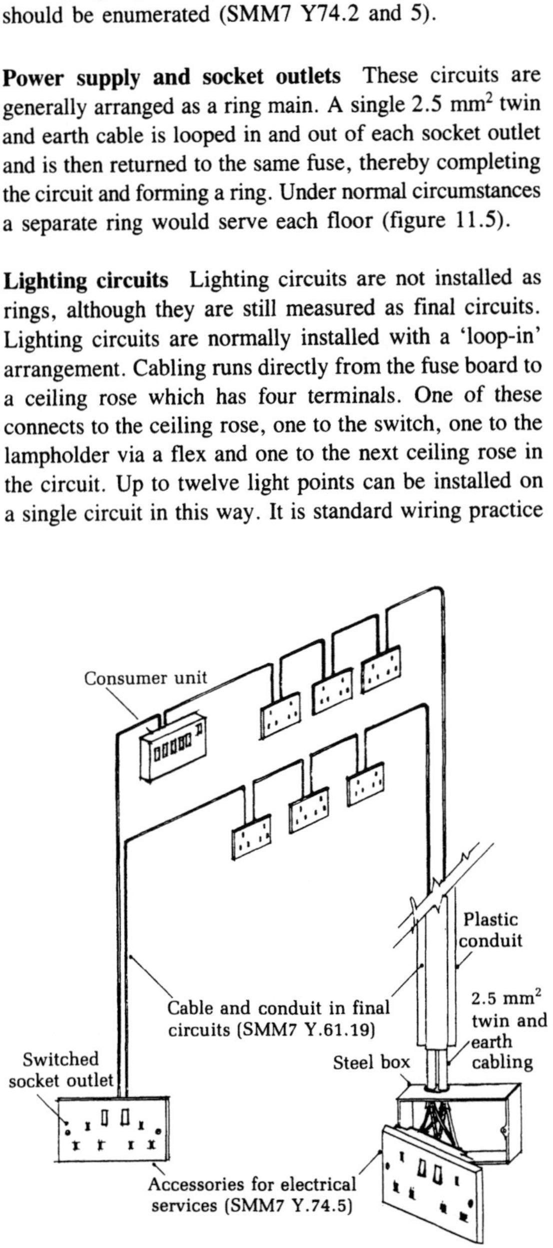

Figure 11.5   Power supply and socket outlets.

to fuse and wire separate circuits for each floor so that the failure of one does not affect the other.

Appliances   Separately fused circuits will be necessary for the cooker, immersion heater, electric heating units and spontaneous water heater or shower.

Work not in final circuits   Mostly found in non-domestic situations where a supply is brought into a premises and then fed to different locations in a building where it is phased down through a fuse-box to supply individual appliances or final circuits (see above). Conduit and cable are each measured separately in linear metres (SMM7 Y60 and Y61).

Concealed wiring and conduit installation   Where cables run behind plastered or dry-lined walling they will be run in conduit. In domestic situations this plastic sleeving prevents the cabling becoming permanently plastered into the wall and allows the cables to be replaced in the event of a fault. Conduit in the final circuit of a domestic installation is included as part of the description accompanying the enumeration of the cabling installation (SMM7 Y61.19.2) (see figure 11.5).

11.3.1 Builder's work in connection

The term 'builder's work in connection' (BWIC) is used to describe a number of operations including cutting holes, forming chases and sinkings, all of which are required to allow the completion of the electrical or plumbing installation. Historically this would have been carried out by the builder's labourer rather than the electrician or plumber; hence the term. The Standard Method includes the measurement of this work under Section P31 and suggests in Measurement Rule M1 that it is included with the measurement of the electrical or plumbing installation. For electrical installations builder's work is measured in accordance with SMM7 P30.19. The description will identify the type of conduit and cabling for each circuit. The number of points served by each individual circuit should be enumerated and described in accordance with SMM7 P30.19.*.1—5. In addition to builder's work in connection it will be necessary to include an item for marking the position of holes, mortices and chases in the structure (SMM7 Y81.2.1). Finally, an inclusion must be made by way of an item for the testing and commissioning of the completed electrical installation (SMM7 Y81.5).

A typical domestic electrical installation might include the measurement items given in table 11.2.

The following take-off relates to no specific example. The descriptions and quantities are typical of those normally encountered in a domestic electrical installation.

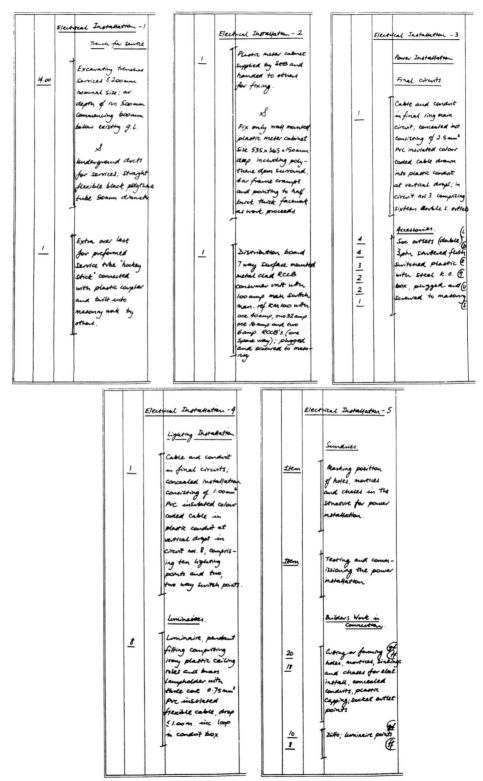

Figure 11.6   Example take-off for electrical installation.

Table 11.2 Domestic electrical installation measurement items

Item	Unit	SMM7 ref.
Excavating trenches for services together with other items associated with excavation work	m	P30.1—9
Underground ducts for engineering services	m	P30.10 and 11
Building in flush electricity meter cabinet. Lintel over, dpc and polythene surround with preformed service tube	nr	F30
Distribution board	nr	Y71.2
Cable and conduit in final circuits:		
switch sockets	nr (of circuits)	Y61.19.2.1
Immersion heaters, etc.	nr (of installations)	Y61.19.2.2
Lighting outlets	nr (of circuits)	Y61.19.2.3
Accessories for elect. inst.:		
luminaires	nr (of fittings)	Y74.2
socket outlets	nr (of outlets)	Y74.5
other fittings	nr (of fittings)	Y74.5
Marking positions of holes and chases etc. for lighting	Item	Y89.2.1
Ditto for sockets	Item	Y89.2.1
Builder's work in connection	Item	P31.19.*.1—5
Testing and commissioning lighting installation	Item	Y89.5.1
Ditto for power installation	Item	Y89.5.1

11.4 Plumbing installation

Plumbing installations in domestic buildings involve a number of different systems associated by the supply, distribution and disposal of water in and about a building. The classification of these is given in Appendix B of the Standard Method and includes the following systems:

R   Disposal Systems
S   Piped Supply Systems
T   Mechanical Heating/Cooling/Refrigeration Systems
U   Ventilation/Air Conditioning Systems

Within each of these there are a number of subsections. (In residential property it is unlikely that any more than a handful will apply.) Work Section Y provides the measurement framework and identifies the components of these systems under a classification which includes pipework, equipment, ducting and insulation.

Pipework is measured in linear metres over all fittings and branches and is deemed to include joints (straight couplings) in the running length. The description should include the type of pipe, its nominal size, the method of jointing and the type, spacing and method of supports. Fittings are enumerated and measured as extra-over the

pipework on which they occur. Where the diameter of the fitting is less than 65 mm the description must make reference to the number of ends (SMM7 Y10.2.3.2—5). For fittings to pipes of diameters larger than 65 mm, the type must be given in the description.

Equipment   In a domestic situation this will include such things as boilers, pumps, cisterns and cylinders. SMM7 Work Sections Y20—Y25 provide the rules for the measurement of general pipeline equipment. These should be enumerated giving in the description the type, size and pattern, rated duty, capacity, loading as appropriate and method of fixing. Since this could include a range of equipment intended to perform different functions, any description is perhaps best cross-referenced to a drawing or specification prepared by the Services Engineer.

The following pages describe the component systems associated with domestic plumbing installation and are followed by a series of take-off lists for the same installation. There is no particular significance intended in the sequence of measurement, other than the need to establish a consistent approach.

Cold water installation   Household water supply enters the property via the Water Company's service pipe (as described earlier in this chapter) and terminates in a stop valve. From this point it rises to a cold water storage cistern usually located in the roof space. A ball valve in the cistern controls the supply of water flowing under pressure from the local service reservoir. In addition to the rising main inlet there are generally a further three pipes connected to the cold water cistern. These are the overflow pipe, the cold water supply to the hot water cylinder and the cold water supply to the bath, wash basin and WC. Pipes are measured in linear metres over all fittings, ignoring any joints (straight couplings) in the running length which are deemed included. In most domestic situations the supply and service pipes are likely to be in copper and this must be given in the description together with the pipe size, the method of jointing (compression or capillary) and the type, spacing and method of fixing. Curved pipework must be classified separately and the radius stated in the description. Fittings such as elbows, tees, reducers and tank connectors together with made bends are enumerated and measured extra-over the pipe on which they occur. Where the diameter is less than 65 mm they are described as fittings with one, two or three ends (copper pipework in a domestic situation would fall into this last category). Made bends, special joints and connectors and fittings with a diameter exceeding 65 mm are also enumerated as extra-over the pipework on which they occur. In the case of joints which differ from those generally occurring in the running length, the description should include the type and method of jointing (SMM7 Y10.2.2.1). Pipework in the roof space

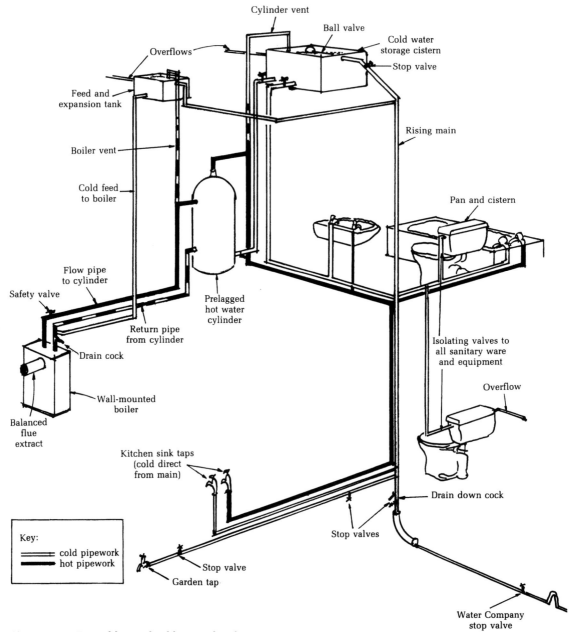

Figure 11.7   Typical hot and cold water distribution system.

should be insulated in accordance with SMM7 Y50.1.1 as should the top and sides of the cold water storage cistern, SMM7 Y50.1.4.2. The measurement of tank stands, platforms and supports should also be included at this stage. Waste pipework, traps and overflow pipework are each measured in similar fashion and are considered further elsewhere.

Hot water installation   This will include the measurement of the boiler (classified under SMM7 T10 and measured under SMM7 Y22.1.1.1), the hot water storage cylinder (SMM7 Y23.1.1.1) and the feed and expansion tank (SMM7 Y21.1.1.1). These should be enumerated, giving in the description the type, size and pattern, rated duty, capacity, loading as appropriate and method of fixing. As previously stated, since this may include a range of equipment intended to perform different functions, any descriptions are perhaps best cross-referenced to a drawing or specification prepared by the Services Engineer. The associated hot water supply pipework would also be included, measured under the same set of rules as cold water pipework.

Heating installation   In a domestic situation where a gas-fired boiler is installed, the hot water and heating installation will share the same heat source. It would be possible to include the measurement of the boiler and hot water cylinder with either the heating or hot water installation using the same set of rules. Likewise, associated pipework should be measured as previously described. Radiators are enumerated as general pipeline equipment (SMM7 Y22.1.1.1), whilst radiator valves, when not supplied with the radiator, are enumerated separately (SMM7 Y22.2.1.1) as ancillaries to the radiator. Often a radiator schedule will be available and descriptions can be extracted from this and presented under an appropriate heading, thereby preventing unnecessary repetition. Most radiators will be delivered to site prefinished; where this is not the case, an item for their decoration must be included in accordance with SMM7 M60.6. The description should state the radiator type (panel or column), whilst the unit of measurement varies depending on the area and girth of the radiator. Pumps, heating controls, motorised valves, immersion heaters and other control equipment associated with the heating system should be included in the same way. Whilst these items should be measured here, it is possible that the cabling and wiring of the control equipment will have been measured with the electrical installation. Care should therefore be taken to ensure that the wiring for this equipment is not included twice.

Sanitary appliances   This includes wash basins, pedestals, baths, bidets, shower trays, WC pans/cisterns and suites. These should be enumerated in accordance with SMM7 N13.4, giving details in the description of their type, size, capacity and method of fixing. Any ancillaries provided with these appliances, including supports and mountings together with the method and background of the fixings, should also be stated. When available full details of the manufacturer, catalogue reference, range and colour should be given. Frequently the cost of supplying sanitaryware is included as a Prime Cost Sum (see chapter 4). In this case separate enumerated fix-only items will need to be measured.

Waste pipework   The rules for the measurement of waste pipework are included with SMM7 Work Section R, Disposal Systems. A distinction is made between foul

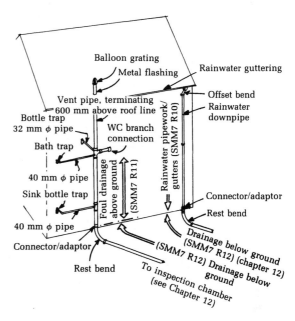

Figure 11.8 Drainage and waste pipework.

drainage installation above and below the ground and the classification for each is given separately. Work below the ground is considered further in the next chapter. Foul drainage pipework above the ground is often referred to as 'waste pipework' and this term is adopted here in the same context. It embraces all the pipework and fittings which are associated with the disposal of used or soiled water. In most modern installations uPVC pipework will discharge into an internal downpipe, known as a stack, or soil and vent pipe, which is in turn connected to the drainage system. The following diameters of pipe are required for the different appliances:

Wash/hand basin	32 mm diameter
Bath/shower/sink	40 mm diameter
WC	110 mm diameter

Overflow pipework (20 or 22 mm diameter) may well be treated and measured under the same set of rules and will be required for WC cisterns, cold water storage cisterns and the feed and expansion cistern.

Waste pipework is measured in linear metres, giving in the description details of the type of pipe, its nominal size, the method of jointing and the spacing and type of pipe brackets. The background and method of fixing must be stated, as should pipework which runs in ducts, chases, floor screeds or in-situ concrete. Straight couplings in the running length of the pipe are deemed included, whilst elbows, tees, tank connectors and access plugs are enumerated as extra-over the pipework on which they occur. Where the diameter of the pipe fitting is less than 65 mm, the description must state the number of connections (SMM7 R11.3.1−4). Where the diameter of pipe fittings exceeds 65 mm, the type of fitting must be stated. Some appliances, such as WC pans, have integral traps, whist others (basins, baths and showers) require separate traps. When measured as separate items these are enumerated as pipework ancillaries in accordance with SMM7 R11.6.8.1, stating the type, method of fixing and nominal size. Testing the waste pipework is measured as an item, giving details in the description of the tests and any attendance required.

In most cases the position of bathroom furniture will be marked on the drawing. Similarly, the stack and the position of the drainage connection should be shown. Given this information it should be possible to measure the waste pipe runs, making appropriate allowances for vertical drops. The stack itself should be carried above the roof line to provide ventilation to the disposal system; alternatively, a pressure release valve may be used in a stub stack installation.

11.4.1 Builder's work in connection

Just as builder's work was included for electrical installation, so it will be measured for mechanical and plumbing work. Holes for pipes through the structure of the building are enumerated, stating the thickness and type of material together with the shape of the hole. These should be classified in accordance with SMM7 P31.20.2.1−3, giving the diameter of the pipe in the following stages: not exceeding 55 mm, 55 mm to 110 mm and exceeding 110 mm. Often, ground-floor service pipework must be laid in ducts in the floor screed. These should be measured in linear metres giving in the description the ducting material, its cross-sectional dimensions, the size and number of services, the depth of screed and the method of fixing. Mortices and sinkings should be enumerated (SMM7 P31.21), whilst chases for pipes should be measured in linear metres (SMM7 P31.22). In each case the description should give the size and number of services together with the nature of the structure. Pipes and ducts sleeved through the structure of the building should be enumerated in accordance with SMM7 P31.23. Associated builder's work items such as bearers, tank stands and pipe casings should also be measured at this stage.

The approach to measuring service installation is very much a matter of individual choice. Some surveyors prefer to follow the flow of water as it enters and is distributed around the building; others choose to measure the 'separate systems' as they are installed, and consequently no specific approach is prescribed. Whichever is adopted, dimensions should be recorded clearly and consistently. The following take-off list relates to no specific example and is intended to be typical of the items that are likely to occur in a domestic installation. It assumes a general approach as follows:

(1) Water service connection
(2) Rising main
(3) Cold water feed
(4) Heating installation
(5) Hot water feed
(6) Sanitaryware
(7) Foul drainage above ground

For most of the above it will be necessary to measure builder's work in connection, marking positions and holes and testing and commissioning.

Take-off List	SMM7 Reference
1. Water Service Connection	
Connection to main	Prov. Sum
Stop valve	Y11.8
Main in trench	Y10.1.1.1.3
Fittings to ditto	Y10.2.3/4
Stop valve and drain tap	Y11.8

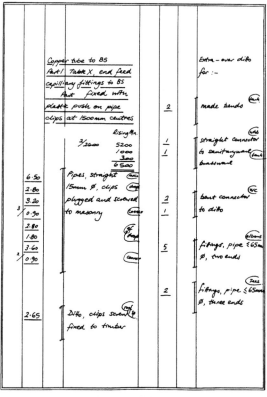

Figure 11.9 Example take-off for domestic plumbing installation and waste pipework.

1	Pipework ancillaries, Connex high pressure gate valve, ref CNX CxC, 2nr. 15mm Ø compression connection to copper tube	
1	Water storage cistern, Perham ref BCJ50, 275 litre capacity including byelaw 30 kit, lid and insulation jacket* size 1200 x 585 x 585mm, holed for one 15mm and three 22mm Ø pipe connections. (measd elsewhere).	
	*Notwithstanding Smm7-Y50.1.4.2 tank ms. measured with equipment	

1	Pipework ancillaries, brass ball valve, high pressure (Part 1); plastic float assembly; connection to copper tube	
3.80 (o'flow)	Pipes, straight 22mm Ø uPVC overflow with solvent welded joints and plastic push on pipe clips at 600mm centres screw fixed to timber	
2	Extra-over ditto for fittings two ends (elbows)	
1	Extra-over for connection to plastic cistern inc. straight tank connector.	

Include the Prime Cost Sum of £1,000 for Sanitary appliances by nom. suppliers

Profit %

Fix only the following Sanitary appliances inc. bedding traps and waste fittings

Vitreous china washdown WC suite inc. 9 litre dual flush cistern with ballvalve silencer, o'flow pipe and plastic float. Matching Celmac seat and cover. overall proj. 705mm height 785mm screw fixing cistern to masonry and pan to timber flooring

1	Vitreous china washbasin and pedestal inc pair brass pillar taps, waste fitting, plug chain and stay. 600 x 495mm proj. x 830mm high inc. plug & screw wall brackets to masonry, screwfix pedestal to timber flooring. & mastic seal basin to wall
1	Reinforced acrylic bath size 1700 x 800mm with pair brass pillar taps, waste fitting, plug, chain and stay, flexible o'flow hose; complete with moulded side panel and screw fixing adjustable feet to timber bearers, panel to timber frame and mastic seal where bath edges abut walls.

Foul drainage above grd.

Soil & Vent pipes

6.50 / 1.80 / 2.40	Pipes, straight uPVC 110mm Ø ring seal joints Osma reference 4S.042/043/044 inc. pipe brackets ref 4S.082 at 2m centres vertically and 1m centres horizontally p.& screw to masonry	

Extra over ditto for:-

1	Balloon grating ref 4S.302 (solvent weld fitting)
2	Offset bend ref 4S.445
1	S/S equal single branch ref 4S.190
1	D/SW Bossed pipe ref 4S.589
1	S/S clayware pipe adaptor ref 4S.107

E.O. ditto (contd)

1	Multi-kwik pan connector ref MK 90S
S	Access bend 90° ref 4S.369
1	Pipework ancillaries vari pitch pipe flashing 450 x 450mm Osma ref 4S.283 (handed to others for fixing) fixing to 110mm Ø uPVC pipework

To take:- waste pipes (s & b) traps (s & b) builders work pipework insulation marking positions testing & commiss. pipework decoration

Heating Installation

Heating Equipment

1	Potterton 'profile 60e' wall mounted balanced flue boiler 80,000 Btu complete with elect. controls, gas burner, control valve, automatic ignition, control gas governor, boiler stat flue and flue terminal, mounting brackets and stove enamelled casing. (2nr) 28mm Ø connections (f&r) and (1nr) 15mm Ø gas connection
1	Grade 3 foam lagged Indirect Hot Water cylinder to BS ___ capacity 110 litres Size 1050 x 400mm diameter, pre-holed for immersion element and 3nr 22mm Ø connections.

To take:- pipework fittings insulation decoration equipment marking pos. test & comm.

Pressed steel double convector Myson Supaline HDC radiators to BS3528 complete with wall brackets p.& s to masonry and hanging radiator

2	480 x 440mm high
2	1440 x 590mm high
1	1760 x 590mm high
5/1	Ancillaries not provided with rads. thermostatic rad. valve 'Honeywell VT200E' 15mm D/Lock Shield compression fitting to 15mm copper tube and PTFE taped fitting to radiator
S	Ditto; radiator valve 'Pegler ternia' 15mm straight; ditto

Figure 11.9 *Continued*

Take-off List	SMM7 Reference
Builder's work in connection:	
Stop cock pit	P30.8
Stop cock box	P30.17
Trench to main	P30.1
Duct for main	P31.10
Fittings to ditto	P31.11.1
2. Rising Main	
Rising main and branches	Y10.1
Fittings to ditto	Y10.2.3/4
Tank connectors	Y10.2.2
Ball valve	Y11.8
Storage cistern	Y21.1
Overflow to cistern	Y10.1
Fittings to ditto	Y10.2.3
Connections	Y10.2.2
Pipe insulation	Y50.1.1.1, Y50.2.1.1
Cistern insulation	Y50.1.4
Builder's work in connection:	
Cistern bearers	G20.13
Tank platform	K20.2.3
Cutting and forming holes	P31.20.2
Cutting mortices and sinkings	P31.21
Cutting chases	P31.22
Pipe casings	G20.12, K20.1.1/2.1
Painting pipes	M60.9.2
Marking position of holes	Y59.1.1
Testing and commissioning	Y51.4
3. Cold Water Feed	
Pipework to taps (cold)	Y10.1
Fittings to ditto	Y10.2.3
Connections to taps and	
cisterns	Y10.2.2
Gatevalves	Y11.8
Insulation and builder's work	
as before	
Marking position of holes	Y51.1
Testing and commissioning	Y51.4
4. Heating Installation	
Boiler	T10 (Y22.1)
Cylinder	Y23.1
Header tank	Y21.1
Insulation to tank	Y50.1.4.2
Immersion heater element	T12 (Y22.1)
Motorised valves	S11 (Y11.1)
Pump	Y20.1
Heating controls	Y53.1
Radiators	Y22.1.1.1
Radiator valves	Y22.2.1.1
Flow and return pipework	Y10.1.1/5
Fittings to same	Y10.2.2/3

Take-off List	SMM7 Reference
Connections	Y10.2.2
Pipework ancillaries:	
Drain cock	Y10.8.1
Safety valve	Y10.8.1
Pipework insulation	Y50.1.1.1, Y50.2.1.1
Builder's work in connection:	
Cistern bearers	G20.13
Tank platform	K20.2.3
Cutting and forming holes	P31.20.2
Cutting mortices and sinkings	P31.21
Cutting chases	P31.22
Pipe casings	G20.12, K20.1.1/2.1
Painting pipes	M60.9.2
Painting radiators	M60.6.1/2.1−3
Marking position of holes	Y59.1.1
Testing and commissioning	Y51.4
Heating pipework:	
Pipework to radiators	Y10.1.1/Y10.1.5
Fittings to ditto	Y10.2.3
Connections to radiators	Y10.2.2
Pipework ancillaries	Y11.8
Pipework insulation	Y50.1.1.1, Y50.2.1.1
Builder's work in connection:	
Cutting and forming holes	P31.20.2
Cutting mortices and sinkings	P31.21
Cutting chases	P31.22
Pipe casings	G20.12, K20.1.1/2.1
Painting pipes	M60.9.2
Marking position of holes	Y59.1.1
Testing and commissioning	Y51.4.1
5. Hot Water Feed	
Pipework to taps (hot)	Y10.1
Fittings to ditto	Y10.2.3
Connections to taps and	
cisterns	Y10.2.2
Pipework ancillaries	Y11.8
Pipework insulation	Y50.1.1.1, Y50.2.1.1
Builder's work in connection:	
Cutting and forming holes	P31.20.2
Cutting mortices and sinkings	P31.21
Cutting chases	P31.22
Pipe casings	G20.12, K20.1.1/2.1
Painting pipes	M60.9.2
Marking position of holes	Y59.1.1
Testing and commissioning	Y51.4.1
6. Sanitaryware	
Wash hand basin	N13.4
Bath	N13.4
WC pan and cistern	N13.4
Bidet	N13.4

Take-off List	SMM7 Reference	Take-off List	SMM7 Reference
Shower tray	N13.4	Fittings to ditto	R11.2.3/4
Builder's work in connection:		Bosses	R11.5.1.1
Cutting and forming holes	P31.20.2	Pipework ancillaries	R11.6.5, R11.6.8
Cutting mortices and sinkings	P31.21	Pipe sleeves	R11.8
Cutting chases	P31.22	Pipework insulation	Y50.1.1.1, Y50.2.1.1
Pipe casings	G20.12, K20.1.1/2.1	Builder's work in connection:	
Painting pipes	M60.9.2	Cutting and forming holes	P31.20.2
Marking position of holes	Y59.1.1	Cutting mortices and sinkings	P31.21
Testing and commissioning	Y51.4.1	Cutting chases	P31.22
		Pipe casings	G20.12, K20.1.1/2.1
7. Foul Drainage Above Ground (wastes, overflows and traps)		Holes in roof tiling	H60.11
		Metal collars	H71.29
Waste pipes/soil pipes/overflow		Painting pipes	M60.9.2
pipes	R11.1	Marking position of holes	Y59.1.1
Connections to ditto	R11.2.2	Testing and commissioning	Y51.4.1

12 Disposal systems below ground

12.1 Drainage below ground

Drainage below ground provides for the dispersal of used and surface water from a building to a point of disposal or treatment. The Standard Method of Measurement distinguishes between disposal systems above and below ground level. Assuming the Group Method of Measurement is adopted, disposal systems above the ground would be measured with their associated work; i.e. rainwater goods with roof coverings (chapter 8) and waste pipework with plumbing installations (chapter 11). SMM7 Work Sections R12 and R13 provide the classification and measurement rules for drainage work below ground level and land drainage disposal systems. Each will require the excavation of trenches, the laying of pipe runs, the provision of beds and the construction of manholes or inspection chambers. For the purposes of measurement it is appropriate to consider the provision of manholes separately from pipe runs. Where separate systems for foul and surface water are proposed, it is sensible to measure each system independently.

The line of drainage runs, together with the position of manholes, will be shown on the drawings. In addition, the drawings should identify each manhole by a reference number (MH1, MH2, etc.) and give details of its cover and invert level. If no reference is given, before measurement can commence it will be necessary to commence at the top of each drainage run and identify each manhole with a specific reference. This is vital since it will provide a unique location code for both schedules and take-off. Where existing and proposed ground levels are provided to coincide with the location of manholes, all to the good; where this is not the case, these will need to be established (by interpolation if necessary). Information regarding ground conditions should be given in a heading (referenced to Section D20.P1) together with details of drainage runs and layout.

A consistent, logical and fully annotated set of dimensions is necessary for this class of work and the preparation of schedules for all but the smallest of systems is recommended. Some surveyors prefer to measure manholes first, whilst others choose to measure drainage runs; either approach is acceptable. The following sequence of measurement has been adopted in the preparation of this chapter:

Foul drainage
(1) Main drainage runs
(2) Branch drainage runs
(3) Pipe fittings and accessories
(4) Manholes/cesspits/septic tanks
(5) Sewer connections
(6) Testing and commissioning

Surface water and land drainage
(1)–(3) As above
(4) Soakaways
(5) Testing and commissioning

12.2 Foul drainage

12.2.1 Main drainage runs

These are measured in linear metres on plan and include the following separate measurements:

- Excavating pipe trenches
- Beds/haunchings/surrounds
- Pipework and fittings

With some care it is frequently possible to ampersand these linear items to a single set of dimensions. Since pipes are built in to the sides of manholes the length of the excavation and the length of the pipe may differ. Where this is the case the inclusion of a separate adjustment to the pipe length will resolve this minor discrepancy.

12.2.2 Manhole and pipe run schedules

Before commencing the take-off for drainage installations the preparation of both pipe run and manhole schedules is recommended for installations of any size. It is perhaps advantageous to prepare the manhole schedule before commencing the pipe run schedule, since some of the information for pipe runs can be abstracted from the manhole schedule (e.g. the depths of pipes will be determined by the invert depths of the manholes). Typically, schedules are prepared on a sheet of abstract paper with the following headings. These are not intended to be inflexible and should be amended to suit the scale and scope of the installation and the individual requirements of the measurer (figures 12.1 and 12.2).

12.2.3 Excavating pipe trenches

Where trenches are to receive pipes of less than 200 mm diameter they are grouped and classified together, stating the average depth of the trench to the nearest 250 mm in the description. Trenches for pipes exceeding 200 mm in diameter are classified separately, stating the nominal pipe size and giving the average depth as before. Pipe trench excavation is deemed to include earthwork support, com-

pacting trench bottoms, trimming excavations, filling with and compaction of filling materials and disposal of surplus excavated material. Where appropriate the description should give the supplementary detail included in SMM7 R12.1.*.*.1−10. Breaking out existing materials and pavings are measured extra-over the previous trench excavation in square metres and cubic metres respectively, stating the type of material in accordance with SMM7 R12.2.1/2.1−5. Where existing hard pavings are to be reinstated this should be stated in accordance with SMM7 R12.2.*.*.1 (refer to Measurement Rule M4 for a definition of trench widths when measuring extra-over items). Excavating pipe trenches alongside existing live services should be measured in linear metres and given extra-over the trench excavation, stating the type of service. Live services which cross the excavation are enumerated as extra-over, giving the same information in the description.

12.2.4 Beds, haunchings and surrounds

Bedding and surround materials for pipes are measured in linear metres and classified in one of the following categories:

(1) Beds

M/H No.	Internal Size Length	Internal Size Width	Ground Level	Inver. Level	Depth to I.L.	BKWK Sides	Slab or Cover	Main Chann.	Conc. Bench	Foot Irons	Remarks
1	1.350	0.778	19.500	17.650	1.850	215 ff o/s	600 × 450 light duty c.i. on 100 mm pcc slab	100 mm φ st.	nil	5 no	All brickwork in class 'B' bricks
2	0.778	0.553	16.750	15.963	0.787	ditto	ditto	100 mm φ curved	1−100 mm φ	nil	ditto
3	0.675	0.450	17.050	16.375	0.675		ditto	100 mm φ st.	2−100 mm φ	nil	ditto
4	0.675	0.450	17.250	16.650	16.600	ditto	ditto	ditto	1−100 mm φ 3/4 sect	nil	ditto

Figure 12.1 Manhole schedule. The headings of this schedule are flexible and can be varied to suit each scheme — it may be desirable to provide a waste column for 'excavation size', 'excavation depth', 'brickwork', etc., which enables all calculations to be carried out on the schedule.

Loc	Length between inside faces of MH	Pipes Type	Pipes Size	Pipes Bends	Length between outside faces of MH	Concrete beds Bed	Concrete beds Haunch	Concrete beds Sur	Length between exc. faces of MH	Depth of dig one end	Depth of dig other end	Av. depth of dig

Figure 12.2 Pipe run schedule.

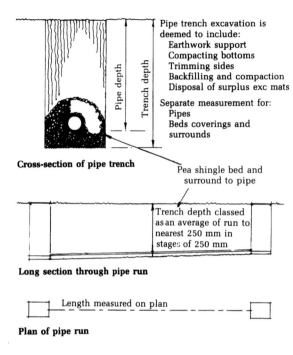

Pipe trench excavation is
deemed to include:
Earthwork support
Compacting bottoms
Trimming sides
Backfilling and compaction
Disposal of surplus exc mats

Separate measurement for:
Pipes
Beds coverings and
surrounds

Pea shingle bed and
surround to pipe

Cross-section of pipe trench

Trench depth classed
as an average of run to
nearest 250 mm in
stages of 250 mm

Long section through pipe run

Length measured on plan

Plan of pipe run

Figure 12.3   Excavation of pipe trenches.

(2)   Beds and haunchings
(3)   Beds and surrounds
(4)   Vertical casings

The width and thickness of beds and haunchings must be given in the description, as should the thickness of any surrounds. The size of vertical pipe casings must also be stated. In all instances, with the exception of (1), the nominal size of the pipe must be given in the description.

12.2.5 Pipework and fittings

Pipes are measured over all fittings and branches in linear metres, stating the kind and quality of pipework and the method of jointing. Pipes in trenches must be so described stating their nominal size; alternative locations are given in SMM7 R12.8.2−4. Pipe fittings, such as bends and branches, are enumerated and measured extra-over the pipe on which they occur. Cutting and jointing pipes to fittings and providing everything necessary for jointing is deemed included. Pipework ancillaries, such as gullies, hoppers, traps and the like, are enumerated, stating the type and giving a dimensioned description which should include the nominal size of inlets and outlets. Jointing ancillaries to pipes and bedding ancillaries in concrete are deemed included.

12.2.6 Manholes and inspection chambers

As with drainage pipework the preparation of a manhole schedule should precede the take-off. Increasingly in domestic situations preformed plastic inspection chambers are used; compared to brick manholes they are easier and cheaper to install. These are usually only permitted for inspection purposes and where the chamber is less than 1.00 metre deep. Where the depth of the manhole exceeds 1.00 m, either traditional brick or precast concrete chamber sections must be used.

The excavation, concretework, formwork, reinforcement, brickwork and rendering associated with manholes must be measured in accordance with the rules for the appropriate Work Section (SMM7 R12 Measurement Rule 8). Building in the ends of pipes, channels, benching, step irons, covers and intercepting traps are each enumerated separately giving a dimensioned description. Note should be made of the different treatment when these form part of a preformed system (SMM7 Measurement Rule 9).

Excavation for manholes is measured in accordance with SMM7 D20.2.4.1−4, in cubic metres, and described as work in pits, giving the number, the depth classification and the commencing level where this exceeds 0.25 m below existing ground level. Disposal items, earthwork support and surface treatments are all measured separately in accordance with SMM7 Section D20. Concrete manhole bases are measured in cubic metres and described as concrete beds, stating the thickness range in stages in accordance with SMM7 E10.4.1−3. Precast concrete cover slabs are enumerated, whilst in-situ slabs are measured in accordance with SMM7 E10.5 with separate items for formwork, reinforcement and cast-in accessories.

Preformed inspection chambers will require the measurement of excavation items and base concrete as previously described. The chamber should be enumerated and described as a preformed system with an accompanying dimensioned description. Building in ends of pipes, channels, benching, step irons and covers are all included as part of the description. Sectionalised precast concrete manholes should be enumerated as separate units, stating the manufacturer and product code, the diameter and length of the chamber sections in the description. Covers, channels, benching and building in ends of pipes should be enumerated separately, accompanied by a dimensioned description.

12.2.7 Connection to mains sewer

In the normal course of events the connection to the main sewer will be carried out by the Local Authority. Where this is the case the cost of this work should be included by way of a Provisional Sum in accordance with SMM7 A53.1. Since it is likely that this will involve the breaking up and later reinstatement of the Local Authority's public highway, the inclusion of a separate Provisional Sum to cover these further costs will also be required. Where this same work is to be carried out by the Contractor, the connection to the Local Authority's sewer should be enumerated in accordance with SMM7 R12.16.1. Separate measurement would also be necessary for trench excavation beyond the boundary of the site, breaking up and later reinstating paved surfaces and any special provision for traffic control and pedestrian safety.

An inclusion by way of an item must also be made for the testing and commissioning of the drainage system, stating the method, purpose and type of installation to be tested. Where attendance or any special instrumentation is required, this should also be given in the description.

12.3 Surface water and land drainage: soakaways

Where it is not permissible or appropriate to discharge rainwater into a sewer, soakaways should be provided. There are specific rules about their construction and location. In domestic situations where ground conditions allow, they are constructed as simple pits filled with granular material. The measurement of excavation items together with the pipe runs associated with soakaways should be carried out using the same set of rules adopted for other drainage work. Filling of soakaway pits should be measured in accordance with SMM7 D20.9.1/2.1−3.

12.4 Worked take-off example

A worked example of a take-off for drainage is shown on the following pages (take-off sheets: pp. 160−67; drawings: pp. 168−70).

DRAINAGE - 2

Information identifying the nature of the subsoil is given in Work Section D20

Details of the layout and installation of the drainage works are all as shown on the accompanying drawing ref. MAF 16

Foul Drainage

main runs. refer schedule (MH1-2)   6.50

Excavating trenches for pipes ≤ 200mm av. depth of trench 750-1000mm   SMM7 R12.1.1.2

Pea gravel beds & surrounds 450mm wide x 10mm bed x 150mm surround to 110mm Ø pipe   SMM7 R12.6.1.1

PVC u piped to BS4660 with pushfit ring seal sockets, in trenches 110mm Ø   SMM7 R12.8.1

Take off list :- (cont.)

(Main & branch drainage runs exc. pipe trench beds pipework fittings accessories)   SURFACE WATER INSTAL.

Soakaways exc. pits disposal brick rubble fill earthwork support adjustment for topsoil

Testing and commissioning

DRAINAGE INSTALLATION - 1

Take off list :-

Main drainage runs - exc. pipe trench beds etc. pipework.

Branch connex. - all as above fittings accessories

Work outside site boundary prov. sums.

Inspection Chambers/MH.
exc. pits
disposal
pea gravel fill
earthwork support
surface treatments
Concrete bed.
preformed I.C. (plastic)
Sectional MH (pcc)
cover slab
main channel
branch channel
benching
b.i. ends of pipes
MH cover & frame

Disposal surface water

Testing & Commissioning

FOUL INSTALLATION

SMM7 R12.8.1.1.3

Vertical pipework measurement

Horizontal pipework msmt

Depth of pipe trench including bed

DRAINAGE — 4

2/ 0.90	PVCu piped all abd in trenches 110mm ∅ vertical	
	& Pea gravel vertical casings 150mm ∅ surround to 110mm ∅ pipe	SMM7 R12.7.1
2/ 1	Extra-over 110mm ∅ PVCu pipe for long radius rest bend Osma code 4D.581	SMM7 R12.9.1.1
1/ 1	Pipe accessories; bottle gully Osma ref 4D.901	SMM7 R12.10.1.1

DRAINAGE — 3

2.50 / 4.30 / 2.80 / 1.00	Excavate trenches MH for pipes all abd at depth of trench 1000-1250mm deep	SMM7 R12.1.1.2
	& Pea shingle bed & surround all abd	SMM7 R12.6.1.1
	PVCu pipe to BS4660 all abd 110mm ∅	SMM7 R12.8.1
1.50 / 1.20 / 2.50	Branch connection MH1 Exc. trench for pipes all abd 1000-1250mm	SMM7 R12.1.1.2
	& Pea gravel beds & surrounds all abd	SMM7 R12.6.1.1
	& Pipes in trenches all abd 110mm ∅	SMM7 R12.8.1

DRAINAGE - 6

MH5 - PCC chamber

Base dims.
length 1350 mm
width 1150 mm   (MH1)

π/	0.39	
	0.39	
	0.87	

Exc. pits (2nr)
max depth
≤1.00m deep   (MH2)

π/	0.39	
	0.39	
	0.98	

Disposal of exc.
mat's off site

Filling to excavation
av. th > 0.25m,
obtained off site,
pea gravel.

π/	0.39	
	0.39	
	1.02	

Exc. pits (3nr)
max depth
≤2.00m deep   (MH3)

π/	0.39	
	0.39	
	1.9	
	1.35	
	1.15	
	1.18	

Disposal of
exc mats off site   (MH4)

Filling to excavation
av. th > 0.25m,
obtained off site,
pea gravel   (MH5)

SMM7 R12 Measurement Rule M8

Excavation, concrete, formwork, brickwork, rendered coatings etc associated with manholes should be measured in accordance with the rules for the appropriate Work Section.

SMM7 D20.2.4.2

SMM7 D20.8.3.1

SMM7 D20.9.2.3

SMM7 D20.2.4.2

SMM7 D20.8.3.1

SMM7 D20.9.2.3

DRAINAGE - 5

Include the following
Provisional Sums for work
carried out by Statutory
Authorities

Prov. Sum £450
for connection to
Local Authority
Sewer

Prov Sum

Prov. Sum £1,500
for breaking up and
later reinstating
public highway

Prov Sum

The following in (4nr)
PVCu preformed inspection
chambers and (1nr)
Sectional PCC manhole

MH1-4

PVCu chamber

450mm

475mm

Exc. diam = 475
backfill π/150   300
Diameter = 775 mm
Radius = 388 mm *
* rounded to nearest mm

SMM7 A53.1.2

Section A Definition Rule D10 includes work carried out by public companies responsible for statutory work.

Work on the public highway can only be carried out by contractors approved by the local Highway Authority and a Road Opening Licence will be needed before work can commence.

DRAINAGE - 7

2/ 1·35 1·18	Earthwork support (MH1) max depth ≤2·00m dist. between opposing faces ≤2·00m
2/ 1·15 1·18	
	SMM7 D20.7.2.1
2/π/ 0·39 0·87	pit face Surf area ex. = 2πr × depth. Earth. support (MH1) max depth ≤1·00m dist. between opposing faces ≤2·00m curved. (MH2)
2/π/ 0·39 0·98	
	SMM7 D20.7.1.1.1
	Circular pits have been measured in accordance with SMM7 General Rule 3.1 — this means Earthwork Support must be measured to the curved exposed faces of the excavation.
2/π/ 0·39 1·02	Ditto, but max (MH3) depth ≤2·00m dist. between opposing faces ≤2·00m; curved. (MH4)
2/π/ 0·39 1·09	
	SMM7 D20.7.2.1.1

DRAINAGE - 8

4/π/ 0·39 0·39	Surface treatmt (MH1) compact bottoms (to4) of exc. & In-situ concrete (20N/mm²) bed thickness ≤ 150mm
	SMM7 D20.13.2.3
1·35 1·15	
	SMM7 E20.4.1
→ × 0·15 = ___ m³	
	Base dimensions converted to volume using constant dimension.
	Osma PVCu underground universal inspection chamber 450mm external diameter all in accordance with BS 4660 and 497
1	Aerformed system comprising chamber base ref 4D.992, chamber shafts ref. 4D.925, Grade C single seal cover and frame ref 4D.344 and building in ends of 2 nr. 110mm Ø pipe (MH1) depth to invert 772mm
	SMM7 R12.12.14.1

DRAINAGE - 10

Milton p.c.c. chambers (cont)

Nr of sections

Excavation dp. 1175
less base 100
bedding 25
cover slab 150 → 275
→ 900

Effective chamber depth = 900
÷; 3nr section

In accordance with SMM7 R12 Measurement Rule 8 p.c.c. manhole chambers have been measured in accordance with SMM7 Work Section E50.

3	Rectangular chamb. Sectional, tongued and grooved edges 305mm deep x 75mm thick walls.

SMM7 E50.1.1.0

1	Cover slab, 1250 x 1050 x 150mm thick with 600x450mm opening.

SMM7 E50.1.1.0

1	Main channel straight ½ round 110mm Ø PVCu Osma rf 4S.874

Since this is not considered a preformed system, items R12.11 to 15.7–13 will require separate measurement.

SMM7 R12.11.8.1

DRAINAGE - 9

Osma P.c.c. underground Universal i.c. (Cont) (MH2)

1	Preformed System comprising; ditto; depth to invert 880mm.

SMM7 R12.12.14.1

1	ditto; bntlang in ends of 3nr 110mm Ø pipes; depth to invert 922mm (MH3)

SMM7 R12.12.14.1

1	ditto; depth to invert 994mm (MH4)

SMM7 R12.12.14.1

Milton p.c.c. chamber sections incl dims. 900 x 700mm with cast in step irons and lifting lugs all in accord with BS 5911 Part 200. joints sealed in c.m. (1:1)

DRAINAGE - 12

Adjustmt for b'fill

MH 1 → 4
ext. diam. 475 mm
ext. radius 238 mm

SMM7 D20.9.2.3

Adjustment for previous over-measurement of backfill (fill material measured to ground level).

π/	0.24	
	0.24	
	0.77	(MH1)

Deduct

Fixing to exc. av. th. >0.25m obtained of site, pea gravel.

π/	0.24	
	0.24	
	0.88	(MH2)
π/	0.24	
	0.24	
	0.92	(MH3)
π/	0.24	
	0.24	
	0.99	(MH4)
	1.05	
	0.85	
	1.05	(MH5)

End of 4(nr) AVcu preformed and (nr) pcc manhole

Item
Disposal surface water
&
Testy & commiss. foul drainage system, water pressure test

SMM7 D20.8.1.0

SMM7 D20.17.1.1

DRAINAGE - 11

1 | Branch channel long-radius bend 3/4 section R.H. 110mm ø PVcu Osma ref 4D788

SMM7 R12.11.8.1

Branch connection measured to MH5 (see drawing) but no inclusion for connecting drainage run.

1 | Benching 900x700mm average 225mm th. in conc (20N/mm²) finished in 9mmo cement mortar (1:3) trowelled smooth to falls and crossfalls

SMM7 R12.11.9.1

3 | Building in ends ø 110mm ø AVcu pipe to sides of 75mm thick pcc chamber sections.

SMM7 R12.11.7.1

1 | Manhole cover and frame BS497 type C. Size 600x450mm setting frame in c.m. (1:3)

SMM7 R12.11.11.1

DRAINAGE - 14

4/0.45	PVCu pipes all abd. in trenches 110mm ø; vertical (rwp Conex)	SMM7 R12.8.1.1.3
	& Pea gravel vertical casings 150mm sand. to 110mm ø pipe	SMM7 R12.7.1
4/1	E.O. 110mm ø PVCu pipe for long radius rt bend Osma code 4D.561 (rwp)	SMM7 R12.9.1.1
	& ditto for 68mm ø rwp connector Osma ref 4D.149	SMM7 R12.9.1.1
2/1	E.O. 110mm ø PVCu pipe for equal junct-ion Osma ref 4D.213 (Ybranch)	SMM7 R12.9.1.1

DRAINAGE - 13

Surface Water

Av depth of pipe rns.
Head = 450
bed 100
= 550 mm

(worst case) 9m rn
@ 1:60 fall = 150mm
∴ depth = 150+550
= 770mm

Av depth classification:
550
770
2)1320
Av = 660mm deep
Classed as 500 - 750mm

9.00 / 5.80 / 5.00	Excavate trench for pipes ≤ 200mm ns av. depth of trench 500 - 750mm (man nu)	SMM7 R12.1.1.2
	& Pea shingle beds 450mm x 100mm deep to 110mm ø pipe (trench nu)	SMM7 R12.6.1.1
1.00 / 3.00	PVCu pipes to BS 4660 with push ft ring seal sockets in trenches 110mm ø	SMM7 R12.8.1

DRAINAGE—16

Soakaways (Cont.)

(pit face surface area)
= 2πr × depth

2 / 2 / π / | 0.75 / 2.50

Earthwork support max depth ≤4.00m dist. between opposing faces ≤2.00m curved.

SMM7 D20.7.3.1

Deduct

2 / π / | 0.75 / 0.75 / 0.15

Filling to exc. earth >0.25m, obtained off site; broken brick rubble

SMM7 D20.9.2.3

Adjustment for 150 mm topsoil cover to rubble filled soakaway.

Add &

Filling to exc. earth ≤0.25m, obtained from on site spoil heaps; topsoil.

SMM7 D20.9.1.1

Assumes topsoil retained on site in temp. spoil heaps.

End of (2nr) brick rubble filled Soakaways

Item.

Testing and commissI surface water drainage system, water test.

SMM7 R12.17.1.1

Disposal surface water item measured with foul drainage work.

DRAINAGE—15

Surface Water (Cont.)

5.40

Pipe accessories: polychannel drainage channel box section 100 wide x 150mm deep with built-in falls and removable metal grating.

SMM7 R12.10.1.1*

*Notwithstanding R12.10.1.1 drainage channels measured in linear metres.

1

Pipe accessories polychannel drain channel all abd Sump box with lift out bucket and 110mm ø outlet.

SMM7 R12.10.1.1

The following in (2nr) brick rubble filled soakaways

Soakaway d=1500 r=0.750

2 / π / | 0.75 / 0.75 / 2.50

Exc. pits (2nr) max depth ≤4.00m

&

Disposal of exc. mats off site.

SMM7 D20.2.4.4

SMM7 D20.8.3.1

&

Filling to exc. earth >0.25m, obtained off site; broken brick rubble

SMM7 D20.9.2.3

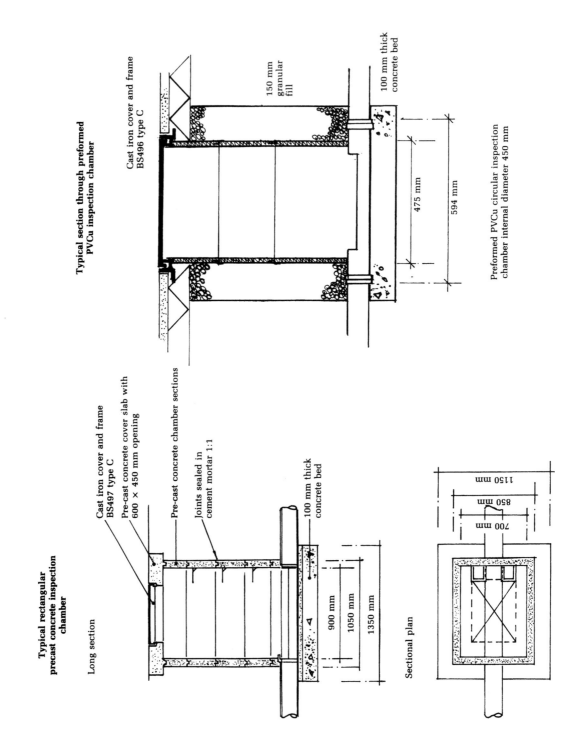

Typical section through preformed
PVCu inspection chamber

Cast iron cover and frame
BS496 type C

150 mm
granular
fill

100 mm thick
concrete bed

475 mm

594 mm

Preformed PVCu circular inspection
chamber internal diameter 450 mm

Typical rectangular
precast concrete inspection
chamber

Long section

Cast iron cover and frame
BS497 type C

Pre-cast concrete cover slab with
600 × 450 mm opening

Pre-cast concrete chamber sections

Joints sealed in
cement mortar 1:1

100 mm thick
concrete bed

900 mm

1050 mm

1350 mm

Sectional plan

1150 mm

850 mm

700 mm

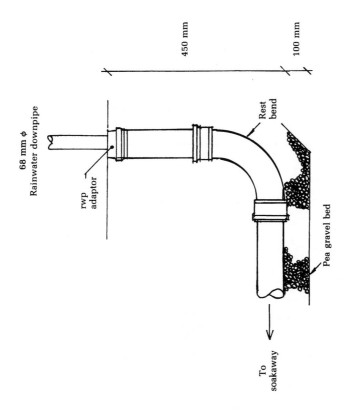

450 mm

100 mm

68 mm φ
Rainwater downpipe

rwp
adaptor

Rest
bend

Pea gravel bed

To
soakaway

**Section through rainwater
downpipe connection**

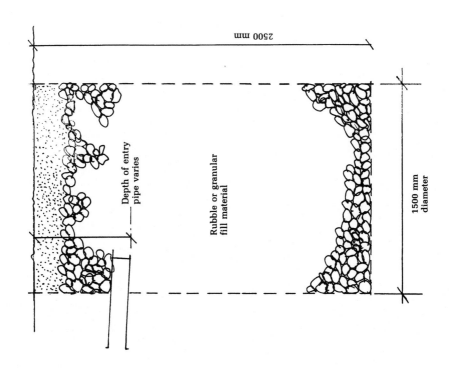

2500 mm

Depth of entry
pipe varies

Rubble or granular
fill material

1500 mm
diameter

Soakaway section

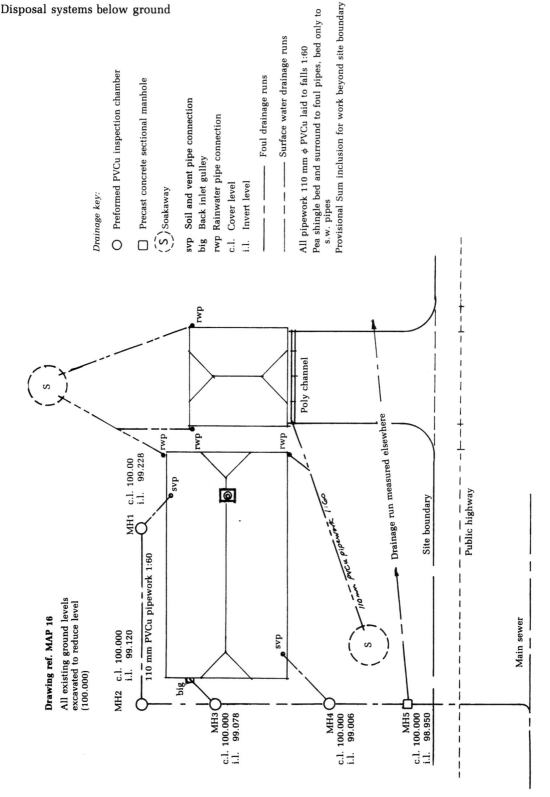

Drawing ref. MAP 16

All existing ground levels
excavated to reduce level
(100.000)

Drainage key:

○ Preformed PVCu inspection chamber

▢ Precast concrete sectional manhole

(S) Soakaway

svp **Soil and vent** pipe connection
big Back inlet gulley
rwp Rainwater pipe connection
c.l. Cover level
i.l. Invert level

———— Foul drainage runs

— · — Surface water drainage runs

All pipework 110 mm φ PVCu laid to falls 1:60
Pea shingle bed and surround to foul pipes, bed only to
s.w. pipes
Provisional Sum inclusion for work beyond site boundary

Use of dimension paper

Each dimension sheet is split into two identical halves, each consisting of four columns:

Column 1 — The *timesing column*, where multiplying numbers may be inserted.

Column 2 — The *dimension column*, where the actual dimensions from the drawings are entered.

Column 3 — The *squaring column*, where the arithmetical products of the dimensions are entered.

Column 4 — The *description column*, where the description of the work associated with the dimensions is entered.

	3·50		**Lineal** Running measurements	
	3·50 2·16		**Superficial** Area measurements	The four forms of dimension
	3·50 2·16 2·00		**Cubic** Volume measurements	Note that for clarity more space should normally be left between forms of measurement than has been shown here
	4		**Enumerated** Numbered items	

Order of dimensions
Should be: length
 breadth or width
 depth or height

| | 10·00 2·00 0·65 | | Concrete in foundations [i.e. 10 m long × 2 m wide × 650 mm deep] |

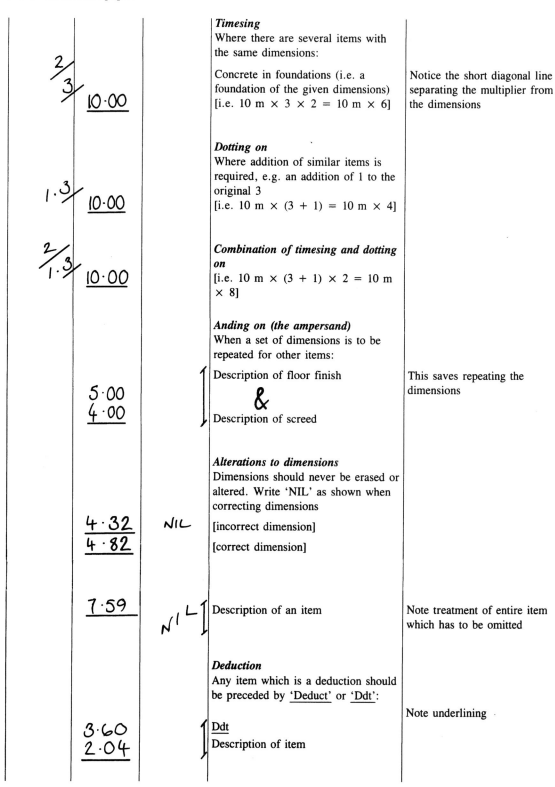

Timesing
Where there are several items with the same dimensions:

Concrete in foundations (i.e. a foundation of the given dimensions) [i.e. 10 m × 3 × 2 = 10 m × 6]

Notice the short diagonal line separating the multiplier from the dimensions

Dotting on
Where addition of similar items is required, e.g. an addition of 1 to the original 3
[i.e. 10 m × (3 + 1) = 10 m × 4]

Combination of timesing and dotting on
[i.e. 10 m × (3 + 1) × 2 = 10 m × 8]

Anding on (the ampersand)
When a set of dimensions is to be repeated for other items:

Description of floor finish

&

Description of screed

This saves repeating the dimensions

Alterations to dimensions
Dimensions should never be erased or altered. Write 'NIL' as shown when correcting dimensions

[incorrect dimension]

[correct dimension]

Description of an item

Note treatment of entire item which has to be omitted

Deduction
Any item which is a deduction should be preceded by 'Deduct' or 'Ddt':

Note underlining

Ddt
Description of item

'Waste' calculations

Dimensions should never be calculated in the head. All preliminary calculations should be made on the dimension paper on the right-hand side of this column to the nearest millimetre:

$$2750$$
$$12000$$
$$3420$$
$$6470$$
$$\overline{24640}$$

All dimensions must be checked and the sources of dimensions always seen

| 24·64 |
| 2·50 |

Description of item

Signposts

Where dimensions can be attributed to locations, such as floors or manholes, notes should be made in this column:

| 43·00 |
| 17·60 |
| 3·85 |

Description of drain pipe   Pipe run

(1 – 2
(2 – 3
(3 – 4

| 8·76 |
| 2·50 |
| 10·24 |
| 2·50 |

Description of paint on walls

(Kitchen

(Bed 1

Numbering and headings

Each dimension sheet should have a heading and be numbered at the bottom of each page:

If shown at top of page denotes 'foundation section' page 6.

Foundations 6

Common abbreviations

a.b.	as before		c.c.	curved cutting
a.b.d.	as before described		C.I.	cast iron
agg.	aggregate		clg.jst.	ceiling joist
a.f.	after fixing		c.jtd.	close jointed
asph.	asphalt		C.P.	chromium plated
av.	average		c.o.e.	curved on elevation
av.g.l.	average ground level		c.o.p.	curved on plan
			c.s.g.	clear sheet glass
b. & p.	bed & point		c.t. & b.	cut tooth and bond
b.e.	both edges		chfd.	chamfered
b.f.	before fixing		chy.	chimney
b.i.	build in		clg.	ceiling
b.m.	birdsmouth		col.	column
b.n.	bull nosed		cos.	course
b.s.	both sides		cpd.	cupboard
bal.	baluster		conc.	concrete
basmt.	basement		csk.	countersunk
bdd.	bedded		cmt.	cement
bdg.	boarding			
bk.	brick		d/d	delivered
bkt.	bracket		Ddt.	deduct
bldg.	building		d.h.	double hung
brd.	board		dia. *or* φ	diameter
brrs.	bearers		dist.	distance
bwk. *or* bkk.	brickwork		D.P.C.	damp proof course
b.o.e.	brick on edge/end		D.P.M.	damp proof membrane
B.S.	British Standard		d.p.	distance piece
			dp.	deep
casmt.	casement		E.G.L.	existing ground level
cav.	cavity		E.M.L.	expanded metal lathing
cav.ins.	cavity insulation		E.O.	extra over
c. & f.	cut and fit		E.S.	earthwork support
c. & p.	cut and pin		ea.	each
c. & s.	cement and sand		exc.	excavate
c.b.	common bricks		excn.	excavation
c.bwk	common brickwork		extl.	external
cc.	centres		extg.	existing

F.A.I.	fresh air inlet	m²	square metre
f.c.	fair cutting	m³	cubic metre
f.f.	fair face	matl.	material
f. & b.	framed and braced	m.g.	make good
f.l. & b.	framed ledged and braced	M.H.	manhole
F.L.	floor level	M.S.	mild steel
fcgs.	facings	m.s.	measured separately
fdns.	foundations	mm	millimetre
fin.	finished	mit.	mitres
fr.	frame	mo.	moulded
frd.	framed	mort.	mortice
fwk.	formwork	msd.	measured
ftd.	fitted		
		n.e.	not exceeding
G.F.	ground floor	Nr.	number
G.I.	galvanised iron		
G.L.	ground level	o/a	overall
galv.	galvanised	o.c.n.	open copper nailing
grano.	granolithic	o.s.	one side
gth.	girth	opg.	opening
		orgl.	original
h.b.s.	herring bone strutting		
h.b.w. or ½ b.w.	half brick wall		
hdb.	hardboard	pbd.	plasterboard
h.c.	hardcore	P.C. sum	Prime Cost sum
hdg.jt.	heading joint	p. & s.	plank and strut
h.m.	hand made	plas.	plaster
hoz.	horizontal	plasd.	pastered
H.P.	high pressure	p.m.	purpose made
h.r.	half round	p.o.	prime only
h.t.	hollow tile	pol.	polished
ht.	height	pr.	pair
hw.	hardwood	Prov. sum	Provisional sum
		prep.	prepare
inc.	including	pt.	point
ins.	insulation	ptd.	pointed
intl.	internal	ptg.	pointing
inv.	invert	ptn.	partition
I.C.	inspection chamber	PVCu	unplasticized polyvinyl chloride
		pvg.	paving
Jap.	Japanned		
jst.	joist	r. & s.	render and set
jt.	joint	r.f. & s.	render float and set
jtd.	jointed	rad.	radius
		R.C.	reinforced concrete
K.P.S.	knot, prime, stop	r.c.	raking cutting
		rdd.	rounded
Lab.	labour	reinf.	reinforced or reinforcement
l. & b.	ledged & braced	R.E.	rodding eye
l.p.	large pipe	R.L.	reduced levels
l. & c.	level and compact	r.l.jt.	red lead joint
		r.m.e.	returned mitred end
m	metre	r.o.j.	rake out joint

R.S.C.	rolled steel channel
R.S.J.	rolled steel joist
R.W.H.	rainwater head
R.W.P.	rainwater pipe
reb.	rebated
retd.	returned
ro.	rough
S.A.A.	satin anodised aluminium
s.b.j.	soldered branch joint
s.d.	screw down
s.c.	stop cock
segtl.	segmental
s.e.	stopped end
s.g.	salt glazed
s.jt.	soldered joint
s.l.	short length
soff.	soffit
s.p.	small pipe
s.q.	small quantities
s.w.	softwood
sk.	sunk
sktg.	skirting
sq.	square
s. & l.	spread and level
S. & V.P.	soil and vent pipe
stg.	starting
swd.	softwood
T.	tee
T. & G.	tongued and grooved
t.	tonne
t. & r.	treads and risers
t.c.	terra cotta
t.p.	turning piece
tops.	topsoil
U.B.	universal beam
uPVC	unplasticized polyvinyl chloride

V.O.	variation order
V.P.	vent pipe
wi. *or* $\overline{w}$	with
w.g.	white glazed
W.l.	wrought iron
W.P.	waste pipe
wdw.	window
wthd.	weathered
X grain	cross grain
X tdg.	cross tongued
mm	millimetre
m	metre
m^2	square metre
m^3	cubic metre
kg	kilogramme

Mathematical symbols

$>$	exceeding
$\geq$	equal to or exceeding
$\leq$	not exceeding
$<$	less than
%	percentage

Example of referencing system used in the Works Section of SMM7

D20:2.2.2.1

D20	Excavation and filling [Work Section]
2	Excavating [number from first column]
2	to reduce levels [number from second column]
2	maximum depth ≤ 1.00 m [number from third column]
1	commencing level stated where >0.25 m below existing ground level [number from fourth column]

Geometric formulae

Title	Figure	Area	Perimeter
Rectangle		lb	$2(l+b)$
Parallelogram		lh	$2(l+b)$
Trapezium		$0.5h(a+b)$	$a+b+c+d$
Triangle		$0.5bh$ $0.5ab \sin C$ $0.5ac \sin B$ $0.5bc \sin A$ $\sqrt{[s(s-a)(s-b)(s-c)]}$ $[s = 0.5(a+b+c)]$	$a+b+c$
Circle		πr^2 $\dfrac{\pi D^2}{4}$	$2\pi r$ πD
Sector		$0.5r^2\theta$ [θ in radians] $0.5rl$ $\dfrac{\pi r^2\theta}{360}$ [θ in radians]	arc length $l = r\theta$ [θ in radians]

Title	Figure	Area	Perimeter
Segment		$0.5r^2(\theta - \sin\theta)$ [θ in radians]	arc length $l = r\theta$ [θ in radians]
Ellipse		πab	$\pi(a+b)$

Title	Figure	Surface area	Volume
Cuboid		$2(ab+bc+ac)$	abc
Pyramid		$(a+b)l+ab$	$\frac{1}{3}abh$
Frustrum of a pyramid		$l(a+b+c+d) + \sqrt{(ab+cd)}$ [regular figure only]	$\dfrac{h}{3}(ab+cd+\sqrt{abcd})$
Wedge		Area ABC'D' = area ABCD $\times$ cos θ area ABCD $= \dfrac{\text{Area ABC'D'}}{\cos\theta}$	area BCC' $\times$ DC
Cylinder		$2\pi rh + 2\pi r^2$	$\pi r^2 h$
Cone		$\pi rl + \pi r^2$ [total surface area]	$\frac{1}{3}\pi r^2 h$
Frustrum of cone		$\pi r^2 + \pi R^2 + \pi(R+r)$ [total surface area]	$\dfrac{\pi h}{3}(R^2+Rr+r^2)$

Title	Figure	Surface area	Volume
Sphere		$4\pi r^2$	$\dfrac{4}{3}\pi r^2$
Segment of sphere		$2\pi Rh$	$\dfrac{\pi h}{6}(3r^2+h^2)$ $\dfrac{\pi h^2}{3}(3R-h)$

Index